Inequalities and Extremal Problems
in Probability and Statistics

Inequalities and Extremal Problems in Probability and Statistics

Selected Topics

Editor and Contributing Author
Iosif Pinelis

Contributing Authors
Victor H. de la Peña

Rustam Ibragimov

Adam Osękowski

Irina Shevtsova

ACADEMIC PRESS

An imprint of Elsevier

Academic Press is an imprint of Elsevier
125 London Wall, London EC2Y 5AS, United Kingdom
525 B Street, Suite 1800, San Diego, CA 92101-4495, United States
50 Hampshire Street, 5th Floor, Cambridge, MA 02139, United States
The Boulevard, Langford Lane, Kidlington, Oxford OX5 1GB, United Kingdom

British Library Cataloguing-in-Publication Data
A catalogue record for this book is available from the British Library

Library of Congress Cataloging-in-Publication Data
A catalog record for this book is available from the Library of Congress

ISBN: 978-0-12-809818-9

For information on all Academic Press publications
visit our website at https://www.elsevier.com/books-and-journals

Working together
to grow libraries in
developing countries

www.elsevier.com • www.bookaid.org

Publisher: Candice Janco
Acquisition Editor: Glyn Jones
Editorial Project Manager: Edward Payne
Production Project Manager: Anusha Sambamoorthy
Cover Designer: Matthew Limbert

Typeset by SPi Global, India

CONTENTS

PREFACE

Inequalities play a fundamental role in mathematics and its many applications. Especially in the statistics and probability literature, there are many more limit theorems than inequalities. However, usually at the heart of a good limit theorem is at least one good inequality. This should become clear if one recalls the definition of the limit and the fact that a neighborhood of a point in a specific topology is usually defined in terms of inequalities. A limit theorem can be very illuminating. However, it only describes the behavior of a function near a given point (possibly at infinity), whereas a corresponding inequality would cover an entire range, oftentimes in many or infinitely many dimensions.

Also, the nature of limit theorems is more qualitative, whereas that of inequalities is more quantitative. For example, a central limit theorem would state that a certain distribution is close to normality; such a statement by itself is qualitative, as it does not specify the degree of closeness under specific conditions. In contrast, a corresponding Berry–Esseen-type inequality can provide quantitative specifics.

This is why good inequalities are important. A good inequality would be, not only broadly enough applicable, but also precise enough; ideally, it would be a solution to an extremal problem. Indeed, such results can be used most effectively in the theory and with a greater degree of confidence and precision in real-world applications. Such an understanding of the role of good and, in particular, best possible bounds goes back at least to Chebyshev. In particular, the theory of Tchebycheff systems was developed to provide optimal solutions to a broad class of such problems. These ideas were further developed by a large number of authors, including Bernstein, Bennett, and Hoeffding. Quoting Bennett (J. Am. Stat. Assoc., 1962):

Much work has been carried out on the asymptotic form of the distribution of such sums [of independent random variables] when the number of component random variables is large and/or when the component variables have identical distributions. The majority of this work, while being suitable for the determination of the asymptotic distribution of sums of random variables, does not provide estimates of the accuracy of such asymptotic distributions when applied to the summation of finite numbers of components. [. . .] Yet, for most practical problems, precisely this distribution function is required.

The contributors to this book are leading experts in the area of inequalities and extremal problems in probability, statistics, and mathematical analysis. It is hoped that the material presented here will promote broader understanding of the importance of inequalities and extremal problems, and that it will stimulate further progress in this area.

The first two chapters of the book, written by Osękowski and devoted to problems arising in the theory of semimartingales, have a strong analytical component. Chapter 1 reviews the so-called method of moments, a powerful and general technique developed in the sixties in the works of Kemperman. The approach, based on dynamic-programming arguments and backward induction, allows the reduction of the study of quite general estimates to the construction of an appropriate functional sequence. This reduction is a common point in many related areas, for example, optimal stochastic control, optimal control theory, and Bellman function method. As an illustration of the method, several new sharp maximal bounds for martingale difference sequences, square function estimates, and prophet inequalities for square-integrable martingales are presented.

Chapter 2 contains a study of a new class of optimal stopping problems for Brownian motion and its maximal function. The classical Markovian approach enables solving such problems of the "integral" form, yielding, in particular, the Doob and Hardy–Littlewood inequalities. The novel method presented in Chapter 2 enables the investigation of optimal stopping problems of "non-integral" type, including Lorentz-norm estimates valid for arbitrary stopping times. The approach rests on inserting an auxiliary optimal stopping problem into the analysis and carrying out an optimization procedure.

Chapter 3, written by Shevtsova, presents the latest and so far the best known universal constants in the so-called nonuniform Berry–Esseen (BE) bounds for sums of independent random variables. Such bounds work better than their uniform counterparts in the tail zones, which are especially important in statistical testing. To a large extent, the method is based on the idea going back to Nagaev and further back to Cramér—to employ the exponential tilt transform to reduce the problem of the nonuniform BE bounds to that of the uniform ones, which latter can be tackled by using, for example, appropriate smoothing inequalities, including ones due to Esseen or Prawitz.

However, this exponential-tilt reduction appears to have inherent limitations, discussed in Chapter 4. Two alternative methods are suggested there, based on novel smoothing inequalities, from which nonuniform BE bounds can be obtained directly, without using the mentioned nonuniform-to-uniform reduction. These new "nonuniform" smoothing inequalities, based in part on fundamental results by Bohman and Prawitz, appear to hold a promise of dramatically improved universal constants in nonuniform BE bounds. As an illustration of how powerful they are, two very short proofs of the classical nonuniform BE bound due to Nagaev are given.

The purpose of Chapter 5 is to establish new uniform BE bounds for the Student statistic or, equivalently, for the so-called self-normalized sums, with explicit constant factors. It is easy to see that the Student statistic is degenerate in the sense that the linear approximation to it is just a rescaled sum of the observations (with a nonrandom scaling factor), and this approximation does not involve squares of the observations. For these reason, it is possible to get a BE bound for the Student statistic involving only the third absolute moments of the observations. However, then the best known associated universal constants are rather large. It is shown in Chapter 5 that BE-type bounds that are substantially better in most practical cases can be obtained if one is allowed to use moments of order higher than 3. Chapters 4 and 5 were contributed by Pinelis.

The concluding chapter (Chapter 6) was written by de la Peña and Ibragimov. Sharp probability and moment inequalities for random polynomials, generalized sample cross-moments, and their self-normalized and Studentized versions, in random variables with an arbitrary dependence are discussed there. The results are based on sharp extensions of probability and moment inequalities for sums of independent random variables to the case of the above statistics in independent symmetric variables. The case of statistics in dependent variables is treated through the use of measures of dependence. The results presented in Chapter 6 are applicable in a number of settings in statistics, econometrics, and time series analysis, including tests for independence and problems of detecting nonlinear dependence.

CHAPTER 1

Method of Moments and Sharp Inequalities for Martingales

Adam Osękowski
University of Warsaw, Warsaw, Poland

1.1 INTRODUCTION

The purpose of this chapter is to introduce the so-called method of moments, an efficient tool which can be used to establish sharp estimates for martingales and other classes of processes. This technique was invented in the sixties in the works of Kemperman (cf. [15, 16]) and applied in several important cases: see, e.g., the works [8–11]. As we will see below, the method rests on a dynamic-programming argument and exploits a certain backward induction. It is also closely related to optimal stopping techniques as well as Burkholder's (or Bellman) method of proving semimartingale inequalities.

Let us introduce the notation which will be used throughout this chapter. Let $(\Omega, \mathcal{F}, \mathbb{P})$ be a probability space, filtered by $(\mathcal{F}_k)_{k \geq 0}$, a nondecreasing family of sub-σ-algebras of $\mathcal{F}$. Let $f = (f_k)_{k=0}^N$ be a finite $(\mathcal{F}_k)_{k \geq 0}$-martingale with the associated difference sequence $df = (df_k)_{k=0}^N$ defined by $df_0 = f_0$ and $df_k = f_k - f_{k-1}$, $k = 1, 2, \ldots, N$. Therefore, we have the equality

$$f_k = df_0 + df_1 + \cdots + df_k, \quad k = 0, 1, 2, \ldots, N.$$

A martingale f is called *simple*, if for each k, the random variable f_k takes only a finite number of values.

Our principal goal is to study sharp inequalities involving f and a certain class of random variables $\mathcal{T}_k(f)$, $k = 0, 1, 2, \ldots, N$, with or without some extra assumptions on the range of these sequences. To be more precise, fix a Borel function $T_0 : \mathbb{R} \to \mathbb{R}$ and a Borel function $T : \mathbb{R}^3 \to \mathbb{R}$. Introduce the sequence $\mathcal{T}(f) = (\mathcal{T}_k(f))_{k=0}^N$ inductively by

$$\mathcal{T}_0(f) = T_0(f_0) \quad \text{and} \quad \mathcal{T}_k(f) = T(f_{k-1}, \mathcal{T}_{k-1}(f), df_k)$$

Inequalities and Extremal Problems in Probability and Statistics. http://dx.doi.org/10.1016/B978-0-12-809818-9.00001-X

for $k = 1, 2, \ldots, N$. A lot of classical objects can be obtained with the use of the above construction. Here are four important examples.

1. *Square function.* Setting $T_0(x) = |x|$ and $T(x, y, z) = (y^2 + z^2)^{1/2}$, we see that

$$\mathcal{T}_n(f) = \left(\sum_{k=0}^{n} df_k^2 \right)^{1/2}, \quad n = 0, 1, 2, \ldots, N$$

is the square function associated with the martingale f.
2. *Maximal functions.* Take $T_0(x) = x$ and $T(x, y, z) = \max\{y, x + z\}$. Then

$$\mathcal{T}_n(f) = \max_{0 \leq k \leq n} f_k, \quad n = 0, 1, 2, \ldots, N$$

is the one-sided maximal function of f. Similarly, the choice $T_0(x) = |x|$ and $T(x, y, z) = \max\{y, |x + z|\}$ corresponds to the two-sided maximal function $\mathcal{T}_n(f) = \max_{0 \leq k \leq n} |f_k|$, $n = 0, 1, 2, \ldots, N$.
3. *Maximal function of the difference sequence.* Take $T_0(x) = |x|$ and $T(x, y, z) = \max\{y, |z|\}$. Then

$$\mathcal{T}_n(f) = \max_{0 \leq k \leq n} |df_k|, \quad n = 0, 1, 2, \ldots, N.$$

4. *Martingale transform.* This example is a slight extension of the above setting, as the transformation T depends on n. Fix a sequence $v = (v_n)_{n=0}^{N}$ of real numbers and put $T_0(x) = v_0 x$, $T_n(x, y, z) = y + v_n z$, $n = 1, 2, \ldots, N$. If we set

$$\mathcal{T}_0(f) = T_0(f_0) \quad \text{and} \quad \mathcal{T}_n(f) = T_n(f_{n-1}, \mathcal{T}_{n-1}(f), df_n)$$

for $n = 1, 2, \ldots, N$, then

$$\mathcal{T}_n(f) = \sum_{k=0}^{n} v_k df_k, \quad n = 0, 1, 2, \ldots, N$$

is the martingale transform of f by the sequence v.

It is easy to see that in many situations we do not need the transformations to be defined on the whole $\mathbb{R}^3$; in general, the pairs $(f_n, \mathcal{T}_n(f))$ take values in some special set $\mathcal{D}$. For example, if $\mathcal{T}(f)$ is the square function of f, then we may take $\mathcal{D} = \mathbb{R} \times [0, \infty)$; if $\mathcal{T}(f)$ is the one-sided maximal function, then $\mathcal{D} = \{(x, y) \in \mathbb{R}^2 : x \leq y\}$; we might consider square function inequalities for nonnegative martingales, and then $\mathcal{D} = [0, \infty) \times [0, \infty)$; and so on. We do not want (and actually not need) to give here a formal definition of $\mathcal{D}$;

instead, we prefer rather to point out that in some situations, for technical reasons, we will work with some special subsets of $\mathbb{R}^2$, those in which $(f, \mathcal{T}(f))$ evolves.

A general problem we will consider can be formulated as follows. Fix the transformations T_0, T as above, suppose that $V : \mathcal{D} \to \mathbb{R}$ is a given function and let N be a fixed nonnegative integer. Assume that we are interested in studying the quantity

$$\inf \, \mathbb{E}V(f_N, \mathcal{T}_N(f)), \tag{1.1.1}$$

where the infimum is taken over all simple martingales $f = (f_k)_{k=0}^{N}$. Here the filtration and the probability space can vary. Note that there are no technical problems with the existence of the above expectations—the integrated random variables take only a finite number of values. The question about the efficient control of the quantity (1.1.1) is of fundamental importance to the theory of martingales and stochastic integration. For example, if for some fixed positive exponent p we set $V(x, y) = |y|^p - C_p^p |x|^p$ and manage to show that (1.1.1) is nonpositive no matter where f starts from, we get the moment bound

$$||\mathcal{T}_N(f)||_{L^p} \leq C_p ||f_N||_{L^p}.$$

Similarly, the choices $V(x, y) = \lambda^p 1_{\{|y| \geq \lambda\}} - c_p |x|^p$, $V(x, y) = |y| - K|x| \log |x| - L$ lead to the corresponding weak-type and logarithmic estimates between $\mathcal{T}_N(f)$ and f_N.

To study the quantity (1.1.1), we consider a more general setting in which the martingale f and the sequence $\mathcal{T}(f)$ can have arbitrary length and start from arbitrary locations. More precisely, introduce the functional sequence $U_n : \mathcal{D} \to \mathbb{R}$, $n = 0, 1, 2, \ldots, N$, by

$$U_n(x, y) = \inf \mathbb{E}V(f_n^x, \mathcal{T}_n^y(f)). \tag{1.1.2}$$

The infimum is taken over the class of all simple martingales $(f_k^x)_{k=0}^{n}$ with $f_0 \equiv x$, and the sequence $\mathcal{T}^y(f) = (\mathcal{T}_k^y(f))_{k=0}^{n}$ is given by $\mathcal{T}_0^y(f) = y$ and $\mathcal{T}_k^y(f) = T(f_{k-1}, \mathcal{T}_{k-1}^y(f), df_k)$, $k = 1, 2, \ldots, n$. Therefore, the sequence $\mathcal{T}^y(f)$ differs from $\mathcal{T}(f)$ only at its initial position: it starts from y instead of $T_0(x)$. The relation of the above problem (1.1.2) to the original setting is evident: the expression in (1.1.1) is equal to $\inf_{x \in \mathbb{R}} U_N(x, T_0(x))$, and hence if we manage to find U_N, we are done.

The reason for considering the whole family $U_0, U_1, \ldots, U_N$ is that it enjoys a nice backward recurrence which in some cases can be solved explicitly. This is described in the following statement.

Theorem 1.1.1. *We have $U_0 = V$. Furthermore, for any $n = 1, 2, \ldots, N$, we have*

$$U_n(x, y) = \inf \ \mathbb{E} U_{n-1}(x + X, T(x, y, X)), \tag{1.1.3}$$

where the infimum is taken over all mean-zero simple random variables X. In particular, $U_n(x, y)$ is the height, at $d = 0$, of the lower boundary of the convex hull of the graph of the function $d \mapsto U_{n-1}(x + d, T(x, y, d))$.

Proof. The equation $U_0 = V$ is trivial. The identity (1.1.3) follows at once by conditioning with respect to the σ-algebra $\mathcal{F}_1$. The geometrical interpretation of $U_n(x, y)$ is an immediate consequence of (1.1.3). □

We conclude this section with several observations.

Remark 1.1.2. It follows directly from Theorem 1.1.1 that if $f = (f_k)_{k=0}^N$ is an arbitrary martingale, then the process $(U_{N-k}(f_k, \mathcal{T}_k(f)))_{k=0}^N$ is a submartingale terminating at the variable $V(f_N, \mathcal{T}_N(f))$. It can be shown that it is actually the largest submartingale bounded from above by the sequence $(V(f_k, \mathcal{T}_k(f)))_{k=0}^N$ (e.g., adapt the reasoning from Chapter 2 in [21]).

Remark 1.1.3. Motivated by the last sentence of the previous remark, we would like to point out that the above approach is very much in the spirit of optimal stopping techniques for Markov processes [22, 24]. Let us explain this precisely, keeping the same notation as in the above martingale setting. Suppose that $f = (f_k)_{k=0}^N$ is a time-homogeneous Markov family on a finite state space $E \subset \mathbb{R}$ with transition function $(p_{ij})_{i,j \in E}$. Let $\mathcal{T}(f)$ be the sequence associated with f and some transformations T_0, T as above, and let $\mathcal{D}$ be the set in which the process $((f_k, \mathcal{T}_k(f)))_{k=0}^N$ takes its values. Fix an arbitrary function $V : \mathcal{D} \to \mathbb{R}$ and study the number

$$\inf \mathbb{E} V(f_\tau, T_\tau(f)), \tag{1.1.4}$$

where the infimum is taken over all stopping times $\tau \leq N$. Note that this can be rewritten as (1.1.1), where that infimum runs over all stopped processes $f^\tau = (f_{\tau \wedge n})_{n=0}^N$ (if only $T_\tau(f) = T_N(f^\tau)$, which is satisfied in most cases).

The classical approach to the study of (1.1.4) is to consider the auxiliary functions

$$U_n(x, y) = \inf \mathbb{E} V(f_\tau^x, T_\tau^y(f)),$$

where the infimum runs over all stopping times $\tau \leq n$ and the sequences f^x, $T^y(f)$ have the same meaning as in the martingale case. Then the sequence $U_0, U_1, \ldots, U_N$ satisfies the recurrence (cf. Chapter I in [22])

$$U_n(x, y) = \min \left\{ \mathbb{E} U_{n-1}(x + X, T(x, y, X)), V(x, y) \right\}, \qquad (1.1.5)$$

where X is the random variable with the distribution given by $\mathbb{P}(X = y - x) = p_{xy}$, $y \in E$. We would like to stress that this is a perfect analogue of (1.1.3) in the Markovian setting. Roughly speaking, both identities come from looking at all possible behaviors of df_1. Indeed, (1.1.3) exploits the fact that the difference df_1 is an arbitrary mean-zero random variable. The identity (1.1.5) comes from the observations that we have only two choices for df_1: either we stop the process (which returns the number $V(x, y)$), or let it evolve according to the transition function (which gives the term $\mathbb{E} U_{n-1}(x + X, T(x, y, X))$).

Remark 1.1.4. As we shall see below, sometimes it is of interest to consider the problem (1.1.1) under some additional assumptions on the range of the terminating variables f_N and $T_N(f)$. Let E be a fixed subset of $\mathbb{R}^2$. Suppose we are interested in the quantity (1.1.1) where the infimum is taken over all martingales $f = (f_k)_{k=0}^N$ such that the pair $(f_N, T_N(f))$ takes values in E. It is easy to see that the above method works also in this setting. Indeed, we define the auxiliary sequence $U_0, U_1, \ldots, U_N$ as above, with the use of (1.1.2), assuming additionally that the terminal variable $(f_n^x, T_n^y(f))$ takes values in E. Clearly, this extra restriction may affect the domain of some of the functions U_n: for instance, U_0 is defined on the set E only. However, it is not difficult to see that the assertion of Theorem 1.1.1 remains valid: the only essential change is that in (1.1.3) one needs to take those mean-zero variables X, for which the pair $(x + X, T(x, y, X))$ falls into the domain of U_{n-1}.

Remark 1.1.5. All the discussion above concerned the martingale setting, but it can be easily adjusted to other related contexts. For example, suppose we are interested in studying the quantity (1.1.1), where the infimum is taken over all simple *submartingales* $(f_k)_{k=0}^n$ (the sequence $T(f)$ is defined in the same manner). Then the analysis goes along the same lines as in the martingale case and requires only some minor modifications. The only essential difference is that in the identity (1.1.3) we need to take the infimum over the class of all simple random variables X satisfying $\mathbb{E}X \geq 0$.

Remark 1.1.6. This remark is motivated by the example 4. concerning martingale transforms. In the above considerations we have assumed "time-homogeneity", that is, that there is only one transformation T which associates with f the corresponding sequence $\mathcal{T}(f)$. However, in some situations for each n there is a whole family of possible transformations to choose from. To be more precise, suppose that $\mathfrak{T}_0$ is a family of some functions $T_0 : \mathbb{R} \to \mathbb{R}$ and let $\mathfrak{T}$ be a family consisting of some functions $T : \mathbb{R}^3 \to \mathbb{R}$. For a martingale f and transformations $T_0 \in \mathfrak{T}_0$, $T_1, T_2, \ldots, T_N \in \mathfrak{T}$, we define

$$\mathcal{T}_0(f) = T_0(f_0) \quad \text{and} \quad \mathcal{T}_n(f) = T_n(f_{n-1}, \mathcal{T}_{n-1}(f), df_n),$$

$n = 1, 2, \ldots, N$. Now one can study the quantity (1.1.1), where the infimum is taken over all simple martingales $f = (f_k)_{k=0}^N$ and all choices of transformations $T_0, T_1, \ldots, T_N$. It is easy to adjust the above approach to this more general setting. Introduce the auxiliary functions $U_0, U_1, \ldots, U_N$ exactly in the same manner as above: $U_n(x, y) = \inf \mathbb{E}V(f_n^x, \mathcal{T}_n^y(f))$, where the infimum runs over all simple martingales $(f_k)_{k=0}^n$ and all transformations $T_0, T_1, \ldots, T_n$. Then the following version of Theorem 1.1.1 holds.

Theorem 1.1.7. *We have $U_0 = V$. Furthermore, for any $n = 1, 2, \ldots, N$, we have*

$$U_n(x, y) = \inf \mathbb{E}U_{n-1}(x + X, T(x, y, X)), \quad (1.1.6)$$

where the infimum is taken over all mean-zero simple random variables X and all transformations $T \in \mathfrak{T}$.

Remark 1.1.8. The final remark links the above technique to the so-called Burkholder's method (or Bellman function method), a powerful tool used widely in probability and analysis to obtain various tight estimates. Namely, in the above considerations we have worked with the finite horizon $0, 1, \ldots, N$, but, obviously, all the questions formulated above have perfect meaning if we let $N = \infty$. There are two essential changes which need to be taken into account: in the definition of the simplicity of the process f, we need to assume that there is a finite deterministic number M such that $f_M = f_{M+1} = f_{M+2} = \cdots$ almost surely (in other words, simplicity implies finiteness); furthermore, for such an f, there must be $\mathcal{T}_M(f) = \mathcal{T}_{M+1}(f) = \mathcal{T}_{M+2}(f) = \cdots$ almost surely (which follows, for instance, from the identity $T(x, y, 0) = y$, satisfied in all the interesting contexts). These two modifications allow to speak about the variable $(f_\infty, \mathcal{T}_\infty(f))$ and hence the quantity (1.1.1) makes sense. Consider the function

$$U_\infty(x, y) = \inf \, \mathbb{E} V(f_\infty, \mathcal{T}_\infty^y(f)),$$

where the infimum is taken over all simple martingales $(f_k)_{k=0}^\infty$ starting from x. Then U_∞ is the pointwise limit of the decreasing sequence U_0, U_1, U_2, ... and hence in particular we have:

1° $U_\infty \leq V$.

Furthermore, letting $n \to \infty$ in (1.1.3) implies the following property
2° For any simple mean-zero random variable X we have

$$U_\infty(x, y) \leq \mathbb{E} U_\infty(x + X, T(x, y, X)).$$

A beautiful feature, exploited by Burkholder in many papers (cf. [3–7]), is that the existence of *a* function U satisfying 1° and 2°—not necessarily equal to U_∞—guarantees the estimate

$$\mathbb{E} V(f_\infty, \mathcal{T}_\infty(f)) \geq U(\mathbb{E} f_\infty, T_0(\mathbb{E} f_\infty))$$

and hence

$$\mathbb{E} V(f_\infty, \mathcal{T}_\infty(f)) \geq \inf_{x \in \mathbb{R}} U(x, T_0(x))$$

for any simple martingale f. Burkholder and his PhD students proved several important martingale inequalities by finding suitable functions U possessing the above properties 1° and 2°. For more on the subject, consult the monograph [21], Suh's paper [25], and Burkholder's works cited earlier.

The remainder of the chapter is devoted to examples: in the next sections we will show how the above method can be used to obtain a number of interesting estimates.

1.2 AN INEQUALITY FOR THE MARTINGALE DIFFERENCE SEQUENCE

In this section we will focus on the following statement, proved by Cox and Kemperman in [10]. Let $df_n^* = \max_{0 \leq k \leq n} |df_k|$, $df^* = \sup_{k \geq 0} |df_k|$ denote the maximal function of the difference sequence.

Theorem 1.2.1. *Let N be a positive integer and let $f = (f_k)_{k=0}^N$ be a martingale. Then*

$$\mathbb{P}(df_N^* \geq 1) \leq \frac{1}{N(2^{1/N} - 1)} \mathbb{E}|f_N|. \tag{1.2.1}$$

The constant $(N(2^{1/N} - 1))^{-1}$ is the best possible: there is a nontrivial martingale f for which both sides are equal.

Letting $N \to \infty$, we immediately obtain the following.

Corollary 1.2.2. *For any martingale $f = (f_k)_{k=0}^\infty$ we have*

$$\mathbb{P}(df^* \geq 1) \leq \frac{1}{\log 2} \sup_{N \geq 1} \mathbb{E}|f_N|$$

and the constant $(\log 2)^{-1}$ is the best possible.

It can be shown that the best constants in (1.2.1) are attained when the left-hand side is equal to 1 (cf. [10]). Furthermore, by a straightforward approximation, we may restrict ourselves to simple martingales. Thus, we will be done if we establish the following fact.

Theorem 1.2.3. *Let N be a positive integer and let $f = (f_k)_{k=0}^N$ be a simple martingale satisfying $df_N^* \geq 1$ almost surely. Then*

$$\mathbb{E}|f_N| \geq N(2^{1/N} - 1) \tag{1.2.2}$$

and the constant on the right can be attained for a nontrivial martingale.

Clearly, the above problem falls into the scope of the method described in the previous section. The random variables (f_n, df_n^*) take values in the set $\mathcal{D} = \mathbb{R} \times [0, \infty)$. Let $V : \mathcal{D} \to \mathbb{R}$ be given by $V(x, y) = |x|$ and let T_0, T be the transformations leading to the maximal function of the difference sequence. We need to show that the quantity in (1.1.1), where the infimum is taken over all martingales for which df_N^* terminates in $[1, \infty)$, is equal to $N(2^{1/N} - 1)$. To handle this problem, we fix $n = 0, 1, 2, \ldots, N$ and introduce the function $U_n : \mathcal{D} \to [0, \infty)$ by (1.1.2), that is,

$$U_n(x, y) = \inf \left\{ \mathbb{E}|f_n| : f_0 \equiv x, y \vee \max_{1 \leq k \leq n} |df_n| \geq 1 \text{ almost surely} \right\}.$$

Using Theorem 1.1.1, we will identify the explicit formulae for U_n. This is the contents of the lemma below.

Lemma 1.2.4.

(i) We have $U_0(x, y) = |x|$, $y \geq 1$; for $y < 1$ the function U_0 is not defined.
(ii) We have

$$U_1(x, y) = \begin{cases} |x| & \text{if } |x| \vee y \geq 1, \\ 1 & \text{if } |x| \vee y < 1. \end{cases}$$

(iii) For $n \geq 2$, we have

$$U_n(x, y) = \begin{cases} |x| & \text{if } |x| \vee y \geq 1, \\ -|x| + 2\left(1 + \dfrac{1 - |x|}{n - 1}\right)^{-n+1} & \text{if } |x| \vee y < 1. \end{cases} \tag{1.2.3}$$

Proof. The first part of the lemma is evident. To study (ii) and (iii), note that $\mathbb{E}|f_n| \geq |\mathbb{E}f_n|$, and hence $U_n(x, y) \geq |x|$ for all x, y and all n. On the other hand, if $|x| \vee y \geq 1$, then directly from the definition of U_n we have the estimate $U_n(x, y) \leq |x|$. Indeed, if $y \geq 1$, then take the constant martingale $f \equiv x$; if $y < 1$, then consider the variable X taking values 1 and -1 with probabilities $1/2$, and take $f_0 \equiv x$, $f_1 = f_2 = \cdots = x + X$. Consequently, for all $n \geq 1$ we may write

$$U_n(x, y) = |x| \quad \text{if } |x| \vee y \geq 1.$$

To get the formula for U_1 on $|x| \vee y < 1$, we apply Theorem 1.1.1. The formula (1.1.3) becomes

$$U_1(x, y) = \inf \{\mathbb{E}|x + X| : X \text{ simple and mean-zero}, \mathbb{P}(|X| \geq 1) = 1\},$$

since we restrict ourselves to martingales satisfying $df_1^* \geq 1$ almost surely. It is evident that the infimum is attained for the Rademacher variable $\mathbb{P}(X = 1) = \mathbb{P}(X = -1) = 1/2$. Indeed: fix a simple mean-zero random variable X satisfying $|X| \geq 1$ almost surely. By the convexity of the function $t \mapsto |x + t|$, the (random) point $(X, |x + X|)$, with probability 1, lies on or above the line passing through $(-1, |x - 1|)$ and $(1, |x + 1|)$. Consequently, the expectation $(\mathbb{E}X, \mathbb{E}|x + X|) = (0, \mathbb{E}|x + X|)$ also has this geometric property, which amounts to saying that $\mathbb{E}|x + X| \geq 1$ (the aforementioned line passes through $(0, 1)$). It remains to note that for the Rademacher variable we have equality; hence

$$U_1(x, y) = 1 \quad \text{if } |x| \vee y \leq 1.$$

We turn our attention to the case $n \geq 2$ and assume that $|x| \vee y < 1$ (for $|x| \vee y \geq 1$ we have already shown the claim). We use Theorem 1.1.1 again and write down (1.1.3):

$$U_n(x, y) = \inf \, \mathbb{E} U_{n-1}(x + X, y \vee |X|),$$

where the infimum is taken over all mean-zero simple random variables X. Note that there are no extra requirements on X due to the restriction $df_n^* \geq 1$ almost surely; this follows at once from the fact that the function U_1, and hence also all the subsequent functions, are given on the full domain $\mathcal{D}$.

So, $U_n(x, y)$ is the height, at $d = 0$, of the boundary of the convex hull of the graph of the function $d \mapsto U_{n-1}(x + d, y \vee |d|)$. The function U_n is symmetric with respect to the variable x (which can be easily shown by induction). Thus, it is enough to show the assertion for $x \geq 0$. Denote the right-hand side of (1.2.3) by $\tilde{U}_n(x, y)$. We will prove that the graph of the function

$$d \mapsto \tilde{U}_n(x, y) + \left(-1 + 2 \left(1 + \frac{1 - x}{n - 1} \right)^{-n+1} \right) d$$

lies below the graph of $d \mapsto U_{n-1}(x + d, y \vee |d|)$ and that both functions coincide for $d = -1$ and $d = (1 - x)/(n - 1)$. This will clearly yield the desired claim.

To show that

$$\tilde{U}_n(x, y) + \left(-1 + 2 \left(1 + \frac{1 - x}{n - 1} \right)^{-n+1} \right) d \leq U_{n-1}(x + d, y \vee |d|),$$

(1.2.4)

suppose first that $n = 2$. The inequality becomes

$$\frac{2}{2 - x}(1 + d) \leq x + d + U_1(x + d, y \vee |d|).$$

We consider several cases. If $d \leq -1$, the left-hand side is nonpositive, while the right is nonnegative (and we have equality for $d = -1$). If $-1 < d < 1 - x$, the inequality is equivalent to $x(x + d - 1) \leq 0$, which is evident. Finally, for $d \geq 1 - x$, some straightforward computations transform the desired bound into $(1 - x)(1 - x - d) \leq 0$, which is trivial (and both sides become equal for $d = 1 - x$).

Now, suppose that $n > 2$. If $d \leq -1$, the inequality (1.2.4) becomes

$$2 \left(1 + \frac{1 - x}{n - 1} \right)^{-n+1} (d + 1) \leq 0,$$

which is obvious (note that for $d = -1$ we get equality). If $-1 < d \leq -x$, (1.2.4) can be rewritten in the form

$$x + d + \left(1 + \frac{1 + x + d}{n - 2} \right)^{-n+2} - \left(1 + \frac{1 - x}{n - 1} \right)^{-n+1} d \geq 0.$$

However, even the sum of the first two terms is nonnegative: this is due to an elementary estimate $u + (1 + u/(n-2))^{-n+2} \geq 1$ valid for $u \geq 0$. Finally, if $d > -x$, we put all the terms of (1.2.4) on the right and denote the obtained expression by $F(d)$. It is easy to check that F is a convex function that vanishes, along with its derivative, at $d = (1 - x)/(n - 1)$. This completes the proof of the lemma. $\qquad\qquad\square$

Proof of Theorem 1.2.3. The above computations show that if f is a martingale as in the statement, then $\mathbb{E}|f_N| \geq U_N(\mathbb{E}f_N, 0)$. However, the function $x \mapsto U_N(x, 0)$ attains its maximal value $N(2^{1/N} - 1)$ at the point $x = 1 - (N - 1)(2^{1/N} - 1)$, and hence (1.2.2) holds true.

It remains to establish the sharpness of the result, which will be obtained by constructing an appropriate example. To do this, it is convenient to use Remark 1.1.2 and rewrite the proof of (1.2.2) just presented above in the form

$$\mathbb{E}|f_N| = \mathbb{E}V(f_N, df_N^*)$$
$$\geq \mathbb{E}U_0(f_N, df_N^*) \geq \cdots \geq \mathbb{E}U_N(f_0, df_0^*) \geq N(2^{1/N} - 1).$$

Let us construct the process for which all the inequalities in the above chain are actually equalities. We start with the equality $\mathbb{E}U_N(f_0, df_0^*) \geq N(2^{1/N} - 1)$: as we have noted above, it will hold if we set $f_0 \equiv 1 - (N - 1)(2^{1/N} - 1)$. Now, we proceed by induction. Suppose that we have constructed the simple martingale $f_0, f_1, \ldots, f_n$ such that $\mathbb{E}U_{N-k}(f_k, df_k^*) = N(2^{1/N} - 1)$ for each $k = 0, 1, \ldots, n$. Suppose that (x, y) is an atom of the variable (f_n, df_n^*). It follows from the above proof that there is a simple, mean-zero random variable X such that

$$U_{N-n}(x, y) = \mathbb{E}U_{N-n-1}(x + X, y \vee |X|).$$

We define f_{n+1} by requiring that conditionally on $\{(f_n, df_n^*) = (x, y)\}$, $f_{n+1} - f_n$ has the same distribution as X. Then by the induction hypothesis, we get

$$\mathbb{E}U_{N-n-1}(f_{n+1}, df_{n+1}^*) = \mathbb{E}U_{N-n}(f_n, df_n^*) = N(2^{1/N} - 1)$$

and we are done.   □

1.3 AN INEQUALITY FOR THE MARTINGALE SQUARE FUNCTION

In this section we will be interested in weak-type bounds for the martingale square function, due to Cox [8]. Here is our main result.

Theorem 1.3.1. *Let N be a positive integer, let $f = (f_k)_{k=0}^N$ be a martingale and let $S_N(f) = \left(\sum_{k=0}^N df_k^2 \right)^{1/2}$ be its square function. Then*

$$\mathbb{P}(S_N(f) \geq 1) \leq \left(1 + N^{-1}\right)^{N/2} \mathbb{E}|f_N| \qquad (1.3.1)$$

and the constant is the best possible: there is a nontrivial martingale for which the equality is attained.

Letting $N \to \infty$, we see that the above result yields the following corollary.

Corollary 1.3.2. *Suppose that $f = (f_k)_{k \geq 0}$ is a real-valued martingale and let $S(f) = \left(\sum_{k=0}^{\infty} df_k^2 \right)^{1/2}$ denote its square function. Then*

$$\mathbb{P}(S(f) \geq 1) \leq e^{1/2} \sup_{N \geq 1} \mathbb{E}|f_N| \qquad (1.3.2)$$

and the constant is the best possible.

It can be shown (cf. [8]) that the constant in (1.3.1) is also optimal when one restricts oneself to martingales f satisfying $\mathbb{P}(S_N(f) \geq 1) = 1$. Furthermore, a straightforward approximation argument proves that it is enough to study simple martingales only. Therefore, it suffices to establish the following statement.

Theorem 1.3.3. *Let N be a positive integer. Then for any martingale $f = (f_k)_{k=0}^N$ satisfying $S_N(f) \geq 1$ almost surely, we have*

$$\mathbb{E}|f_N| \geq \left(1 + N^{-1}\right)^{-N/2}. \qquad (1.3.3)$$

The above theorem can be studied by the method described in Section 1.1. We take $\mathcal{D} = \mathbb{R} \times [0, \infty)$ and define the function $V : \mathcal{D} \to \mathbb{R}$ by $V(x, y) = |x|$. Furthermore, we pick the transformations T_0, T leading to the square function. For $n \geq 0$, let $U_n : \mathcal{D} \to \mathbb{R}$ be given by (1.1.2), which for the above choice of parameters becomes

$$U_n(x, y) = \inf \left\{ \mathbb{E}|f_n| : f_0 = x, y^2 + \sum_{k=1}^{n} df_k^2 \geq 1 \text{ almost surely} \right\}$$

$$= \inf \left\{ \mathbb{E}|f_n| : f_0 = x, y^2 - x^2 + S_n^2(f) \geq 1 \text{ almost surely} \right\}.$$

Lemma 1.3.4. *The functions U_n have the following explicit formulae.*

(i) If $y \geq 1$, then $U_0(x, y) = |x|$; for $y < 1$, the function U_0 is not defined.
(ii) We have

$$U_1(x, y) = \begin{cases} |x| & \text{if } x^2 + y^2 \geq 1, \\ \sqrt{1 - y^2} & \text{if } x^2 + y^2 < 1. \end{cases}$$

(iii) For $n \geq 2$,

$$U_n(x, y)$$
$$= \begin{cases} (n-1)^{(n-1)/2}(1-y^2)^{n/2}(n-x^2-ny^2)^{(1-n)/2} & \text{if } x^2 + y^2 < 1, \\ |x| & \text{if } x^2 + y^2 \geq 1. \end{cases}$$

Proof. The first part is evident. To prove (ii) and (iii), observe that $\mathbb{E}|f_n| \geq |\mathbb{E}f_n|$, so $U_n(x, y) \geq |x|$ for all n, x, and y. On the other hand, if $x^2 + y^2 \geq 1$, then $U_n(x, y) \leq |x|$, directly from the definition of U_n. Indeed, if $y \geq 1$, it suffices to consider the constant martingale $f \equiv x$, while for $y < 1$, one takes the martingale given by $f_0 \equiv x$ and $f_1 = f_2 = \cdots = f_n$, where f_1 takes values $|x| \pm \sqrt{1 - y^2}$ with probability $1/2$. The assumption $x^2 + y^2 \geq 1$ guarantees that f does not change its sign and therefore $\mathbb{E}|f_n| = |x|$ for all n. Consequently, we have

$$U_n(x, y) = |x| \quad \text{if } x^2 + y^2 \geq 1.$$

To find the formula for U_n on $x^2 + y^2 < 1$, consider first the case $n = 1$. Then the definition of the function becomes

$$U_1(x, y)$$
$$= \inf \left\{ \mathbb{E}|x + X| : X \text{ simple and mean-zero}, \mathbb{P}(X^2 + y^2 \geq 1) = 1 \right\},$$

where the requirement $X^2 + y^2 \geq 1$ comes from the fact that $(x + X, (y^2 + X^2)^{1/2})$ must belong to the domain of U_0. It is easy to see that the above infimum is attained for X having the distribution $\mathbb{P}(X = -\sqrt{1 - y^2}) = \mathbb{P}(X = \sqrt{1 - y^2}) = 1/2$: other choices for X return larger values of the expectation (the reasoning is similar to that presented in the proof of Lemma 1.2.4 and we will omit it). So, we have

$$U_1(x, y) = \sqrt{1 - y^2} \quad \text{if } x^2 + y^2 < 1.$$

To find the formula for U_n when $n \geq 2$, observe that (1.1.3) yields

$$U_n(x, y) = \inf \mathbb{E} U_{n-1}\left(x + X, (y^2 + X^2)^{1/2}\right),$$

where the infimum is taken over all mean-zero random variables X (no extra assumptions on the range of X are needed, the domain of U_n, $n \geq 1$, is a full halfplane $\mathcal{D}$). Consequently, $U_n(x, y)$ is the height, at $d = 0$, of the lower boundary of the convex hull of the graph of the function $d \mapsto U_{n-1}\left(x + d, (y^2 + d^2)^{1/2}\right)$. We have already shown the assertion for $x^2 + y^2 \geq 1$, so we may assume that $x^2 + y^2 < 1$. We use induction. We will show that there is a coefficient $A = A_n(x, y)$ such that the graph of the linear function $d \mapsto \tilde{U}_n(x, y) + Ad$ lies below the graph of $d \mapsto U_{n-1}\left(x + d, (y^2 + d^2)^{1/2}\right)$, and both functions are equal for some $d_- < 0$ and $d_+ > 0$. (Here $\tilde{U}_n$ is the function given by the expression on the right-hand side of the equation in (iii)). This will give the claim.

We start with the case $n = 2$ and set $A = x(2 - 2y^2 - x^2)^{-1/2}$. Consider the inequality

$$\tilde{U}_2(x, y) + Ad \leq U_1\left(x + d, (y^2 + d^2)^{1/2}\right).$$

One easily checks that for $d_\pm = (-x \pm (2 - 2y^2 - x^2)^{1/2})/2$ both sides are equal (note that $d_- < 0$ and $d_+ > 0$, since $x^2 + y^2 < 1$). Furthermore, we have $|A| < 1$, the function $d \mapsto U_1\left(x + d, (y^2 + d^2)^{1/2}\right)$ is continuous, linear on $(-\infty, d_-]$ and $[d_+, \infty)$ (with the corresponding slopes -1 and 1, respectively), and concave on $[d_-, d_+]$. See Fig. 1.1. This shows the claim for $n = 2$.

The case $n > 2$ is much more elaborate (cf. [8]). One can show that for fixed x and y, there is a unique A such that

$$\tilde{U}_n(x, y) + Ad \leq U_{n-1}\left(x + d, (y^2 + d^2)^{1/2}\right) \tag{1.3.4}$$

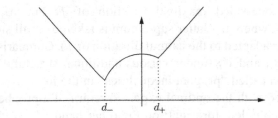

Fig. 1.1 *The graph of the function* $d \mapsto U_1 \left(x + d, (y^2 + d^2)^{1/2} \right)$ *for* $x = 0.5$ *and* $y = 0.1$.

holds for all $d \in \mathbb{R}$. Let us sketch the proof of this fact. Let $d_\pm$ be positive and negative solutions to the equation

$$(x + d)^2 = \frac{1 + (n - 2)x^2 - y^2}{n + (n - 2)x^2 - ny^2}(1 + x^2 - y^2 + 2xd).$$

First one proves that there is A such that (1.3.4) becomes an equality for both $d_\pm$. Having done this, we introduce the function F : $d \mapsto U_{n-1}\left(x + d, (y^2 + d^2)^{1/2}\right)$ and check that we have equality $F'(d_-) = F'(d_+) = A$; finally, one shows that there are numbers $c_\pm$ with $d_- < c_- < c_+ < d_+$ such that the function is convex on each of the intervals $(-\infty, c_-], [c_+, \infty)$ and concave on $[c_-, c_+]$. We omit the tedious verification of these properties. This gives (1.3.4) and completes the proof of the lemma.   □

Proof of Theorem 1.3.3. We have shown above that for each N and any martingale f as in the statement we have the inequality

$$\mathbb{E}|f_N| \geq U_N(\mathbb{E}f_N, |\mathbb{E}f_N|).$$

Therefore, it suffices to observe that the function $x \mapsto U_N(x, |x|)$ attains its maximal value $(1+N^{-1})^{-N/2}$ at $x = (N+1)^{-1/2}$. This completes the proof: the sharpness is dealt with exactly in the same manner as in the previous section.   □

1.4 A PROPHET INEQUALITY FOR L^2-BOUNDED MARTINGALES

In this section we establish another interesting estimate for finite martingales [19], which belongs to the class of the so-called prophet inequalities. To give the reader some motivation behind this class, assume that $f = (f_0, f_1, f_2, \ldots, f_n)$ is a sequence of random variables and let $f_n^* = \max_{0 \leq k \leq n} f_k$

stand for the one-sided maximal function of f. Put $M_n = \mathbb{E}f_n^*$ and $V_n = \sup_\tau \mathbb{E}f_\tau$, where the latter supremum is taken over all stopping times τ of f (i.e., all τ adapted to the natural filtration of f). Comparisons between the numbers M_n and V_n (under various additional structural assumptions on f) have been called "prophet inequalities" in the literature. Clearly, M_n can be identified with the optimal expected return of a prophet or a player endowed with complete foresight; on the other hand, V_n can be treated as an optimal expected return of a player who knows only past and present, but not the future. Prophet inequalities have played a distinguished role in the theory of optimal stopping and have been studied intensively in the eighties and nineties. We refer the interested reader to the works [1, 12–14, 17, 18, 23] and consult references therein.

We will apply the method of Section 1.1 to establish the following result. Throughout this section, f^* stands for the one-sided maximal function of f: $f_n^* = \max_{0 \le k \le n} f_k$, $n = 0, 1, 2, \ldots$.

Theorem 1.4.1. *Let the sequence $k = (k_n)_{n \ge 0}$ be given by the conditions $k_0 = 0$ and*

$$k_n = (1 + k_{n-1}^2)/2, \quad n \ge 1.$$

Then for each $N \ge 1$ and any L^2-bounded martingale $f = (f_k)_{k=0}^N$ we have the inequality

$$\mathbb{E}f_N^* \le \mathbb{E}f_N + k_N \sqrt{\operatorname{Var} f_N}. \tag{1.4.1}$$

The inequality is sharp.

It suffices to study the above inequality for simple martingales only. Let $T_0(x) = x$, $T(x, y, z) = \max\{x + z, y\}$ be the transformations corresponding to the one-sided maximal function and set $\mathcal{D} = \{(x, y) \in \mathbb{R}^2 : x \le y\}$. At the first glance, the inequality (1.4.1) cannot be translated into the study of an expression of the form (1.1.1). To overcome this difficulty, we will apply an extra homogenization argument. Namely, let us first consider an auxiliary problem with $V : \mathcal{D} \to \mathbb{R}$ given by $V(x, y) = x^2 - y$. For any nonnegative integer n and any $(x, y) \in \mathcal{D}$, let $U_n(x, y)$ be given by (1.1.2). That is,

$$U_n(x, y) = \inf \mathbb{E}\left[f_n^2 - y \vee \max_{1 \le k \le n} f_k\right],$$

where the infimum is taken over all martingales $f = (f_k)_{k=0}^n$ satisfying $f_0 = x$ almost surely. In contrast to the previous estimates, we do not assume anything about the range of the terminal variable $(f_n, y \vee \max_{1 \le k \le n} f_k)$.

An application of (1.1.3) yields

$$U_n(x,y) = \inf \{\mathbb{E}U_{n-1}(x+X, (x+X) \vee y) : X \text{ simple and mean-zero}\} \tag{1.4.2}$$

for $n = 1, 2, \ldots, N$. We turn to the explicit formula for U_n.

Lemma 1.4.2. *For any* $n = 0, 1, 2, \ldots$ *we have*

$$U_n(x,y) = \begin{cases} (2y - k_n)x - y - (y - k_n/2)^2 & \text{if } y - x < k_n/2, \\ x^2 - y & \text{if } y - x \ge k_n/2. \end{cases} \tag{1.4.3}$$

Proof. We proceed using induction. If $n = 0$, then the identity (1.4.3) holds true, since $k_0 = 0$ and $U_0 = V$. Suppose that (1.4.3) is valid for some nonnegative $n-1$ and let us try to compute U_n with the use of (1.4.2). To this end, introduce the function $h : \mathbb{R} \to \mathbb{R}$ by $h(d) = U_{n-1}(x+d, (x+d) \vee y)$. A direct computation shows that $h(d)$ is given by the formula

$$\begin{cases} (x+d)^2 - y & \text{if } d \le y - x - k_{n-1}/2, \\ (2y - k_{n-1})(x+d) - (y - k_{n-1}/2)^2 - y & \text{if } y - x - k_{n-1}/2 < d \le y - x, \\ (x+d)^2 - (x+d) - k_{n-1}^2/4 & \text{if } d > y - x. \end{cases}$$

Let us describe the convex hull of the graph of h. We easily check that

- h is continuous,
- h is convex and of class C^1 on each of the intervals $(-\infty, y - x)$, $(y - x, \infty)$,
- its one-sided derivatives at $d = y - x$ satisfy $h'(y - x-) \ge h'(y - x+)$,
- h is linear on $(y - x - k_{n-1}/2, y - x)$.

See Fig. 1.2.

Thus, we need to find a common tangent line to the parabolas $\gamma_1 : d \mapsto (x+d)^2 - y$ and $\gamma_2 : d \mapsto (x+d)^2 - (x+d) - k_{n-1}^2/4$. A little calculation gives that this line is

$$\{(s,t) : t = (2y - k_n)(x+s) - y - (y - k_n/2)^2\},$$

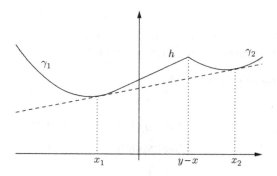

Fig. 1.2 *The common tangent to the parabolas γ_1 and γ_2 lies below the graph of h.*

and the tangency points are

$$(x_1, \gamma_1(x_1)) = \left(y - x - k_n/2, \gamma_1(y - x - k_n/2)\right),$$
$$(x_2, \gamma_2(x_2)) = \left(y - x - k_n/2 + 1/2, \gamma_2(y - x - k_n/2 + 1/2)\right).$$

Thus, by the graphical interpretation of U_n and the fact that x_2 is nonnegative, we obtain that $U_n(x, y) = x^2 - y$ if $x_1 \geq 0$ and $U_n(x, y) = (2y - k_n)x - y - (x - k_n/2)^2$ if $x_1 < 0$. This is precisely the claim. □

Proof of (1.4.1). Consider the centered martingale $\tilde{f} = (f_0 - \mathbb{E}f_0, f_1 - \mathbb{E}f_0, f_2 - \mathbb{E}f_0, \ldots, f_N - \mathbb{E}f_0)$. Applying the definition of U_N conditionally with respect to $\mathcal{F}_0$, we get

$$\mathbb{E}\left(\tilde{f}_N^2 - \tilde{f}_N^*\right) \geq \mathbb{E}U_N(\tilde{f}_0, \tilde{f}_0) = \mathbb{E}\tilde{f}_0^2 - \mathbb{E}\tilde{f}_0 - k_N^2/4.$$

However, $\tilde{f}_0$ has expectation 0, so the latter expression is not smaller than $-k_N^2/4$. Consequently,

$$\mathbb{E}f_N^* - \mathbb{E}f_0 = \mathbb{E}\tilde{f}_N^* \leq \mathbb{E}\tilde{f}_N^2 + k_N^2/4 = \mathrm{Var}\, f_N + k_N^2/4.$$

Applying this inequality to the rescaled martingale f/λ (where λ is a fixed positive constant), we obtain

$$\mathbb{E}f_N^* - \mathbb{E}f_0 \leq \lambda^{-1}\mathrm{Var}\, f_N + \lambda k_N^2/4.$$

The right-hand side, as a function of λ, attains its minimum for the choice $\lambda = 2\left(\mathrm{Var}\, f_N\right)^{1/2}/k_N$. Plugging this value of λ above, we obtain the desired estimate (1.4.1). The sharpness is established in a similar manner as in Section 1.2. □

1.5 MOMENT INEQUALITY FOR MARTINGALE SQUARE FUNCTION

The method studied above has the drawback of computational complexity (we have already experienced it when studying the weak-type bound for the martingale square function). However, in some cases the computations can be simplified, by considering a slight modification of the sequence U_0, $U_1, U_2, \ldots$. The purpose of this section is to illustrate this phenomenon, by looking at the following statement.

Theorem 1.5.1. *Let $(C_n)_{n \geq 0}$ be the sequence of numbers given by $C_0 = 1$ and $C_{n+1} = 1 + C_n^2/4$, $n = 0, 1, 2, \ldots$. Then for any nonnegative integer N and any martingale $f = (f_k)_{k=0}^N$ we have*

$$\|f_N\|_1 \leq C_N \|S_N(f)\|_1. \tag{1.5.1}$$

The constant C_N is the best possible.

This result can be studied with the use of the technique described in Section 1.1, by setting $V(x, y) = C_N y - |x|$, $(x, y) \in \mathcal{D} = \mathbb{R} \times [0, \infty)$, and letting T_0, T be the transformations leading to the square function. However, there seems to be no simple formulae for the associated functions $U_0, U_1, \ldots, U_N$. To overcome this difficulty, we exploit Remark 1.1.2: if $f = (f_k)_{k=0}^N$ is a martingale starting from x, then

$$
\begin{aligned}
U_N(x, |x|) &= \mathbb{E}U_N\left(f_0, S_0(f)\right) \\
&\leq \mathbb{E}U_{N-1}\left(f_1, S_1(f)\right) \\
&\leq \cdots \\
&\leq \mathbb{E}U_0\left(f_N, S_N(f)\right) \\
&= \mathbb{E}V\left(f_N, S_N(f)\right).
\end{aligned}
\tag{1.5.2}
$$

The idea is that one may search for other functional sequences (in the place of $(U_k)_{k=0}^N$), for which the above chain of inequalities holds true (in the last line, we allow the bound "$\leq$", instead of equality).

Let us state this observation separately.

Theorem 1.5.2. *Let $V : \mathcal{D} \to \mathbb{R}$ be a given function and let N be a fixed nonnegative integer. Suppose that $(U_k)_{k=0}^N$ is a sequence of real-valued functions on $\mathcal{D}$, which satisfies the following three conditions:*

(i) We have $U_N(x, |x|) \geq 0$ for all $x \in \mathbb{R}$.
(ii) We have $U_0(x, y) \leq V(x, y)$ for all $x \in \mathbb{R}$ and $y \geq 0$.

(iii) For each $k = 1, 2, \ldots, N$ there is a function $A_k : \mathbb{R} \times [0, \infty) \to \mathbb{R}$ such that the following holds: if $x, d \in \mathbb{R}$ and $y \geq 0$, then

$$U_k(x, y) + A_k(x, y)d \leq U_{k-1}\left(x + d, (y^2 + d^2)^{1/2}\right).$$

Then the inequality $\mathbb{E}V(f_N, S_N(f)) \geq 0$ holds true for all martingales $(f_k)_{k=0}^N$.

The proof is immediate and rests on checking that the chain (1.5.2) is valid. We omit the details.

In the proof of Theorem 1.5.1, we will need the following technical lemmas. For the proofs, we refer the interested reader to [20]. We assume that $C \in [1, 2]$ is a fixed parameter.

Lemma 1.5.3. *For any $x, d \in \mathbb{R}$ such that $|x| \leq \sqrt{C - 1}$ we have*

$$|x + d| - \sqrt{C}\sqrt{x^2 + 1 + Cd^2} \leq \sqrt{C - 1}(-1 + xd).$$

Lemma 1.5.4. *For any $y \geq 0$, $d \in \mathbb{R}$, and $x \geq \sqrt{C - 1}y$ we have*

$$|x + d| - C\sqrt{y^2 + d^2} \leq x - 2\sqrt{C - 1}y + (C - 1)d.$$

Lemma 1.5.5. *Assume that the numbers $x, y \geq 0$, and $d \in \mathbb{R}$ satisfy the conditions $x \geq \sqrt{C - 1}y$ and $|x + d| \leq \frac{C}{2}\sqrt{y^2 + d^2}$. Then*

$$-\frac{C}{2}\sqrt{\left(\frac{C^2}{4} + 1\right)(y^2 + d^2) - (x + d)^2} - (C - 1)d \leq x - 2\sqrt{C - 1}y.$$

$$(1.5.3)$$

Equipped with the above statements, we are ready to introduce a family of special functions; the author obtained these objects as the result of an elaborate experimentation. For a fixed $1 \leq C \leq 2$, let $U^C : \mathbb{R} \times [0, \infty) \to \mathbb{R}$ be given by

$$U^C(x, y) = \begin{cases} \frac{C}{2}\sqrt{(\frac{C^2}{4} + 1)y^2 - x^2} & \text{if} \quad |x| \leq \frac{C}{2}y, \\ -|x| + Cy & \text{if} \quad |x| > \frac{C}{2}y. \end{cases} \qquad (1.5.4)$$

We will also need an auxiliary function A^C on $\mathbb{R} \times [0, \infty)$, defined by

$$A^C(x, y) = \begin{cases} -\frac{C}{2}x / \sqrt{(\frac{C^2}{4} + 1)y^2 - x^2} & \text{if} \quad |x| \leq \frac{C}{2}y, \\ -\frac{C^2}{4}\operatorname{sgn} x & \text{if} \quad |x| > \frac{C}{2}y. \end{cases}$$

Let $(C_n)_{n\geq 0}$ be the sequence introduced in the statement of Theorem 1.5.1 and let $N \geq 0$ be fixed. For any $k = 0, 1, 2, \ldots, N$, let $U_k = U^{C_{N-k}}$ and $A_k = A^{C_{N-k}}$. Finally, put $V(x,y) = C_N y - |x|$. We will show that the sequence $(U_k)_{k=1}^N$ has all the necessary properties listed in Theorem 1.5.2.

Lemma 1.5.6. *The conditions (i) and (ii) of Theorem 1.5.2 are satisfied.*

Proof. The property (i) is trivial: we have $U_N(x, |x|) = U^{C_0}(x, |x|) = 0$. The condition (ii) also has a simple proof. Indeed, for $|x| \geq C_N y/2$ we get equality, so we may assume that $|x| < C_N y/2$. Furthermore, we may restrict ourselves to nonnegative x. Rewrite the majorization in the form

$$C_N y \geq x + \frac{C_N}{2}\sqrt{\left(\frac{C_N^2}{4} + 1\right)y^2 - x^2} \qquad (1.5.5)$$

and observe that the right-hand side, as a function of $x \in [0, C_N y/2]$, is increasing: its derivative equals

$$1 - \frac{C_N}{2}\frac{1}{\sqrt{(\frac{C_N^2}{4}+1)\frac{y^2}{x^2} - 1}} \geq 0.$$

Hence, it suffices to show (1.5.5) for $x = C_N y/2$; but then both sides are equal. $\qquad\square$

We turn to the third condition of Theorem 1.5.2.

Lemma 1.5.7. *For any $k = 1, 2, \ldots, N$, any $x, d \in \mathbb{R}$ and any $y \geq 0$ we have*

$$U_{k-1}\left(x + d, \sqrt{y^2 + d^2}\right) \geq U_k(x,y) + A_k(x,y)d. \qquad (1.5.6)$$

Proof. Denote $C = C_{N+1-k}$, so that $C_{N-k} = 2\sqrt{C-1}$. The function U_{k-1} is defined by the right-hand side of (1.5.4), while the formulas for U_k and A_k read

$$U_k(x,y) = \begin{cases} -\sqrt{C-1}\sqrt{Cy^2 - x^2} & \text{if} \quad |x| \leq \sqrt{C-1}y, \\ |x| - 2\sqrt{C-1}y & \text{if} \quad |x| > \sqrt{C-1}y. \end{cases}$$

and

$$A_k(x,y) = \begin{cases} \sqrt{C-1}x/\sqrt{Cy^2 - x^2} & \text{if} \quad |x| \leq \sqrt{C-1}y, \\ (C-1)\,\text{sgn}\,x & \text{if} \quad |x| > \sqrt{C-1}y. \end{cases}$$

Suppose first that $x \leq \sqrt{C-1}y$. If $|x+d| \leq \frac{C}{2}\sqrt{y^2+d^2}$, then

$$U_{k-1}(x+d, \sqrt{y^2+d^2}) = -\frac{C}{2}\sqrt{\left(\frac{C^2}{4}+1\right)(y^2+d^2)-(x+d)^2}$$

$$\leq -\sqrt{C-1}\sqrt{C(y^2+d^2)-(x+d)^2}$$

(simply square both sides to verify the latter bound). The discriminant of the quadratic function $d \mapsto C(y^2+d^2) - (x+d)^2$ is nonpositive (because of the assumption $x \leq \sqrt{C-1}y$), so the function $H(d) = -\sqrt{C-1}\sqrt{C(y^2+d^2)-(x+d)^2}$ is concave. Thus $H(d) \leq H(0) + H'(0)d$, or

$$U_{k-1}(x+d, \sqrt{y^2+d^2}) \leq -\sqrt{C-1}\sqrt{Cy^2-x^2} + \sqrt{C-1}\frac{xd}{\sqrt{Cy^2-x^2}},$$

which is precisely (1.5.6).

Next, assume that $x \leq \sqrt{C-1}y$ and $|x+d| > \frac{C}{2}\sqrt{y^2+d^2}$. The bound (1.5.6) becomes

$$|x+d| - C\sqrt{y^2+d^2} \leq \sqrt{C-1}\sqrt{Cy^2-x^2}\left(-1+\frac{xd}{Cy^2-x^2}\right).$$

By homogeneity, we may assume that $Cy^2 - x^2 = 1$. Then the above inequality is precisely the assertion of Lemma 1.5.3. Therefore, all we need is the verification of the assumption $|x| \leq \sqrt{C-1}$ appearing in the statement of the lemma. But this follows from

$$x^2 = Cx^2 - (C-1)x^2 \leq C(C-1)y^2 - (C-1)x^2 = C-1.$$

Next, we turn to the case $|x| \geq \sqrt{C-1}y$. Since $U_k(x, y) = U_k(-x, y)$ and $A_k(x, y) = -A_k(-x, y)$, we may restrict ourselves to nonnegative x. Now, if $|x+d| > \frac{C}{2}\sqrt{y^2+d^2}$, the inequality (1.5.6) is precisely the assertion of Lemma 1.5.4. On the other hand, if $|x+d| \leq \frac{C}{2}\sqrt{y^2+d^2}$, then the claim follows from Lemma 1.5.5. $\square$

The two lemmas above yield the validity of (1.5.1). In contrast to the preceding estimates, it can be shown that equality holds for trivial martingales only. For the detailed proof of the sharpness, we refer the reader to [20].

1.6 AN INEQUALITY FOR MARTINGALE TRANSFORMS

The final example we will study here is the following weak-type bound for martingale transforms.

Theorem 1.6.1. *Suppose that N is a positive integer. Then for any martingale $f = (f_k)_{k=0}^{N}$ and any deterministic sequence $v = (v_k)_{k=0}^{N}$ with values in $\{-1, 1\}$,*

$$\mathbb{P}\left(\left|\sum_{k=0}^{N} v_k df_k\right| \geq 1\right) \leq 2\mathbb{E}|f_N|.$$

The constant is the best possible.

This result was originally established by Burkholder in [2], slightly different proofs can be found in [3, 4]. It turns out that the optimal constant is attained for martingales f such that the left-hand side is 1 (cf. [3]). Therefore, it is enough to establish the following statement, to study which Remark 1.1.4 will be used.

Theorem 1.6.2. *Suppose that N is a positive integer and let $f = (f_k)_{k=0}^{N}$ be a martingale such that*

$$\mathbb{P}\left(\left|\sum_{k=0}^{N} v_k df_k\right| \geq 1\right) = 1$$

for some deterministic sequence $v = (v_k)_{k=0}^{N}$ with values in $\{-1, 1\}$. Then we have

$$\mathbb{E}|f_N| \geq 1/2, \tag{1.6.1}$$

and the lower bound cannot be improved.

It is easy to see that we may restrict ourselves to simple martingales, and then we are in the situation described in Remark 1.1.6. For each n, we have two transformations to choose from (corresponding to the choice of $v_n \in \{-1, 1\}$). That is, we take $\mathfrak{T}_0 = \{T_{0-}, T_{0+}\}$, where $T_{0\pm}(x) = \pm\, x$, and, for $n \geq 1$, we set $\mathfrak{T}_n = \{T_-, T_+\}$, where $T_{\pm}(x, y, z) = y \pm z$. Let $V : \mathcal{D} = \mathbb{R} \times \mathbb{R} \to \mathbb{R}$ be given by $V(x, y) = |x|$ and define the associated sequence U_n by

$$U_n(x, y) = \inf\, \mathbb{E}|f_n|,$$

where the infimum runs over all martingales $(f_k)_{k=0}^n$ starting from x such that

$$\mathbb{P}\left(\left|y + \sum_{k=1}^n v_k df_k\right| \geq 1\right) = 1$$

for some deterministic sequence $v = (v_k)_{k=1}^n$ with values in $\{-1, 1\}$. The identity (1.1.6) implies that for any $n \geq 1$ we have

$$U_n(x, y) = \inf \ \mathbb{E}U_{n-1}(x + X, T_\pm(x, y, X)), \qquad (1.6.2)$$

where the infimum is taken over all simple, mean-zero random variables X and all choices $T_\pm$ for the transformation.

Lemma 1.6.3. *The functions $U_0, U_1, \ldots, U_N$ admit the following explicit formulae.*

(i) We have $U_0(x, y) = |x|$ if $|y| \geq 1$; for $|y| < 1$ the function is not defined.
(ii) We have

$$U_1(x, y) = \begin{cases} 1 - y^2 - |xy| & \text{if } |x| + |y| < 1, \\ |x| & \text{if } |x| + |y| \geq 1. \end{cases}$$

(iii) For $n \geq 2$,

$$U_n(x, y) = \begin{cases} (x^2 - y^2 + 1)/2 & \text{if } |x| + |y| < 1, \\ |x| & \text{if } |x| + |y| \geq 1. \end{cases}$$

Proof. The formula for U_0 is evident. To prove the second and the third part of the lemma, observe that $\mathbb{E}|f_n| \geq |\mathbb{E}f_n|$ and hence $U_n(x, y) \geq |x|$ for all n, x, and y. On the other hand, if $|x| + |y| \geq 1$, then there is a simple, mean-zero random variable X and a number $\varepsilon \in \{-1, 1\}$ such that $x + X$ has the same sign as x and $|y + \varepsilon X| \geq 1$. To see this, suppose first that $|y| \geq 1$; then the random variable $X \equiv 0$ has all the required properties. If $|y| < 1$ and $|x| + |y| \geq 1$, then we consider four cases, depending on the signs of x and y. If, say, $x, y \geq 0$, then we put $\varepsilon = -1$ and consider X taking values in the set $-x, 1 + y$; the remaining three possibilities are handled similarly. Plugging the martingale $(x, x + X, x + X, \ldots, x + X)$ into the definition of $U_n(x, y)$, we see that $U_n(x, y) \leq |x|$ and thus $U_n(x, y) = |x|$ for $n \geq 1$ and $|x| + |y| \geq 1$.

Therefore, it remains to check the formula for $U_n(x, y)$ for $|x| + |y| < 1$. If $n = 1$, (1.6.2) can be rewritten in the form

$$U_1(x, y) = \min\{U_1^+(x, y), U_1^-(x, y)\}. \qquad (1.6.3)$$

Here $U_1^{\pm}(x,y)$ are given by

$$U_1^{\pm}(x,y) = \inf\{\mathbb{E}|x+X|\}$$

and the infimum is taken over all simple centered random variables X such that $|y \pm X| \geq 1$ almost surely. Now, the infimum defining $U_1^+(x,y)$ is attained for the random variable X with the distribution $\mathbb{P}(X = -y+1) = (1+y)/2$, $\mathbb{P}(X = -y-1) = (1-y)/2$ (modify appropriately the argument used above in the proof of Lemma 1.2.4), and hence

$$U_1^+(x,y) = \frac{1+y}{2}|x-y+1| + \frac{1-y}{2}|x-y-1| = 1 + xy - y^2.$$

For $U_1^-(x,y)$ we proceed analogously and obtain $U_1^-(x,y) = 1 - xy - y^2$; combining this with (1.6.3) yields the second part of the lemma.

If $n=2$, we use the formula (1.6.2) again. One easily checks that for $|x| + |y| < 1$ and $d \in \mathbb{R}$ we have

$$\frac{x^2 - y^2 + 1}{2} + xd \mp yd \leq U_1(x+d, y \pm d). \qquad (1.6.4)$$

To show this, note we may take $\mp = -$ and $\pm = +$, replacing y with $-y$, if necessary. If $|x+d| + |y+d| \leq 1$, the inequality becomes

$$\frac{x^2 - y^2 + 1}{2} + xd - yd \leq 1 - (y+d)^2 - |(x+d)(y+d)|,$$

which is equivalent to $(|x+d| + |y+d|)^2 \leq 1$. If $|x+d| + |y+d| > 1$, we must show that

$$\frac{x^2 - y^2 + 1}{2} + xd - yd \leq |x+d|,$$

or $||x+d| - 1| \leq |y+d|$. By assumption, $|y+d| \geq 1 - |x+d|$; furthermore,

$$|y+d| \geq |d| - |y| \geq |d| + |x| - 1 \geq |x+d| - 1.$$

This gives (1.6.4), which in turn, in the light of (1.6.2), implies $U_2(x,y) \geq (x^2 - y^2 + 1)/2$ (for $|x| + |y| < 1$). To see that we actually have equality here, we consider an appropriate example. If $d \in \{(1-x-y)/2, (x+y-1)/2\}$, then $|x+d| + |y+d| = 1$ almost surely and hence both sides of (1.6.4) are equal (again, with the choice $\mp = -, \pm = +$). Therefore, if X is a mean-zero random variable taking values in the set $\{(1-x-y)/2, (x+y-1)/2\}$, then

$$\frac{x^2 - y^2 + 1}{2} = \mathbb{E}U_1(x+X, y+X)$$

and hence the assertion for $n = 2$ is proved. To show that the functions U_3, U_4, ... are equal to U_2, it suffices to verify that the function U_2 is concave along the lines of slope ± 1 (we omit the straightforward calculation). Therefore, when applying (1.6.2), we see that the optimal choice for X is $X \equiv 0$ (and any transformation $T_\pm$). $\qquad\square$

Proof of Theorem 1.6.2. It is easy to see that for any $N \geq 1$ and any $x \in \mathbb{R}$ we have $U_N(x, \pm x) \geq 1/2$; this proves the validity of (1.6.1). To show that the bound is optimal, we may argue as in Sections 1.2–1.4, but we can also provide a simple example. Namely, the nonnegative martingale f starting from $1/2$ and satisfying $f_1 = f_2 = \cdots = 1/2 + X$, where $\mathbb{P}(X = -1/2) = 3/4$ and $\mathbb{P}(X = 3/2) = 1/4$, enjoys $\mathbb{E}|f_N| = 1/2$ and $|df_0 - df_1 - df_2 - df_3 - \cdots - df_N| = 1$ almost surely. $\qquad\square$

REFERENCES

[1] D. Assaf, E. Samuel-Cahn, Simple ratio prophet inequalities for a mortal with multiple choices, J. Appl. Probab. 37 (4) (2000) 1084–1091.

[2] D.L. Burkholder, A sharp inequality for martingale transforms, Ann. Probab. 7 (1979) 858–863.

[3] D.L. Burkholder, Boundary value problems and sharp inequalities for martingale transforms, Ann. Probab. 12 (1984) 647–702.

[4] D.L. Burkholder, Explorations in martingale theory and its applications, in: Ecole d'Eté de Probabilités de Saint Flour XIX-1989, Lecture Notes in Mathematics, vol. 1464, 1991, pp. 1–66.

[5] D.L. Burkholder, Strong differential subordination and stochastic integration, Ann. Probab. 22 (1994) 995–1025.

[6] D.L. Burkholder, Sharp norm comparison of martingale maximal functions and stochastic integrals, in: Proceedings of the Norbert Wiener Centenary Congress 1994 (East Lansing, MI, 1994), Proc. Symp. Appl. Math., vol. 52, American Mathematical Society, Providence, RI, 1997, pp. 343–358.

[7] D.L. Burkholder, The best constant in the Davis inequality for the expectation of the martingale square function, Trans. Am. Math. Soc. 354 (2002) 91–105.

[8] D.C. Cox, The best constant in Burkholder's weak-L^1 inequality for the martingale square function, Proc. Am. Math. Soc. 85 (1982) 427–433.

[9] D.C. Cox, Some sharp martingale inequalities related to Doob's inequality, in: Inequalities in Statistics and Probability (Lincoln, Neb., 1982), IMS Lecture Notes Monogr. Ser., vol. 5, Institute of Mathematical Statistics, Hayward, CA, 1984, pp. 78–83.

[10] D.C. Cox, J.H.B. Kemperman, On a class of martingale inequalities, J. Multivar. Anal. 13 (1983) 328–352.

[11] D.C. Cox, R.P. Kertz, Common strict character of some sharp infinite-sequence martingale inequalities, Stoch. Process. Appl. 20 (1) (1985) 169–179.

[12] T.P. Hill, R.P. Kertz, Additive comparisons of stop rule and supremum expectations of uniformly bounded independent random variables, Proc. Am. Math. Soc. 83 (1981) 582–585.

[13] T.P. Hill, R.P. Kertz, Stop rule inequalities for uniformly bounded sequences of random variables, Trans. Am. Math. Soc. 278 (1983) 197–207.

[14] T.P. Hill, R.P. Kertz, A survey of prophet inequalities in optimal stopping theory, in: Strategies for Sequential Search and Selection in Real Time (Amherst, MA, 1990), Contemp. Math., vol. 125, American Mathematical Society, Providence, RI, 1992, pp. 191–207.

[15] J.H.B. Kemperman, The general moment problem, a geometric approach, Ann. Math. Stat. 39 (1968) 93–122.

[16] J.H.B. Kemperman, Geometry of the moment problem, in: Moments in Mathematics, Proc. Symp. Appl. Math., vol. 37, American Mathematical Society, Providence, RI, 1987, pp. 16–53.

[17] D.P. Kennedy, R.P. Kertz, A prophet inequality for independent random variables with finite variances, J. Appl. Probab. 34 (1997) 945–958.

[18] H. Kösters, Difference prophet inequalities for [0, 1]-valued i.i.d. random variables with cost for observations, Ann. Probab. 32 (4) (2004) 3324–3332.

[19] A. Osękowski, A prophet inequality for L^2-bounded martingales, Stat. Probab. Lett. 83 (10) (2013) 2319–2323.

[20] A. Osękowski, Moment inequality for the martingale square function, Bull. Pol. Acad. Sci. Math. 61 (2) (2013) 169–180.

[21] A. Osękowski, Sharp Martingale and Semimartingale Inequalities, in: Monografie Matematyczne, vol. 72, Birkhäuser, Basel, 2012, 462 pp.

[22] G. Peskir, A.N. Shiryaev, Optimal Stopping and Free Boundary Problems, in: Lecture Notes in Mathematics, ETH Zurich, Zurich, 2006.

[23] E. Samuel-Cahn, A difference prophet inequality for bounded i.i.d. variables, with cost for observations, Ann. Probab. 20 (3) (1992) 1222–1228.

[24] A.N. Shiryaev, Statistical Sequential Analysis. Optimal Stopping Rules, second ed. (revised), Izdat. "Nauka", Moscow, 1976, 272 pp.

[25] Y. Suh, A sharp weak type (p, p) inequality $(p > 2)$ for martingale transforms and other subordinate martingales, Trans. Am. Math. Soc. 357 (4) (2005) 1545–1564.

CHAPTER 2

On a Class of Optimal Stopping Problems of Nonintegral Type

Adam Osękowski
University of Warsaw, Warsaw, Poland

2.1 INTRODUCTION

The purpose of this chapter is to study a class of "nonintegral" optimal stopping problems for Brownian maximal functions. This type of problems, being very natural and interesting, does not fall into scope of general methodology and its analysis requires the development of novel ideas. Though we will focus on the maximal setting only, we strongly believe that the arguments can be carried over to a much wider setting, for instance for a more general class of diffusions. This seems to be a wide area awaiting further research.

We begin by providing the main setup in which we work. Suppose that $(\Omega, \mathcal{F}, \mathbb{P})$ is a complete probability space, equipped with a filtration $(\mathcal{F}_t)_{t\geq 0}$. Let $B = (B_t)_{t\geq 0}$ be an adapted, standard Brownian motion starting from 0 and denote by $B^* = (B_t^*)_{t\geq 0}$ its (two-sided) maximal function, given by $B_t^* = \sup_{0\leq s\leq t} |B_s|$. We will be interested in comparing the sizes of B and B^* stopped at a certain adapted Markov time τ. More precisely, introduce the set

$$E = \{(x, y) \in \mathbb{R} \times [0, \infty) : y \geq |x|\}, \qquad (2.1.1)$$

the state space of the process (B, B^*). Suppose that $G : E \to \mathbb{R}$ is a fixed Borel function (called *gain function*) and assume that we are interested in showing the inequality

$$\mathbb{E}G(B_\tau, B_\tau^*) \leq 0 \qquad (2.1.2)$$

for all stopping times τ (typically, one restricts oneself to stopping times satisfying certain boundedness properties depending on G, guaranteeing that the above expectation exists). This problem can be treated by means of optimal stopping theory, by considering the quantity

Inequalities and Extremal Problems in Probability and Statistics. http://dx.doi.org/10.1016/B978-0-12-809818-9.00002-1

29

$$\sup \mathbb{E} G(B_\tau, B_\tau^*), \tag{2.1.3}$$

the supremum being taken over all stopping times τ as above. There is a well-known general methodology to study problems of this type. Roughly speaking, one extends the problem (2.1.3) to the case in which the process (B, B^*) starts at an arbitrary point $(x, y) \in E$. Then, denoting the corresponding supremum in (2.1.3) by $U(x, y)$, one exploits Markovian arguments and shows that the obtained *value function* $U : E \to \mathbb{R} \cup \{\infty\}$ enjoys certain structural properties (it satisfies an appropriate majorization and a partial differential equation on a part of its domain). These conditions enable an explicit identification of U, and (2.1.2) follows once one checks that $U(0, 0) \leq 0$. A more detailed description of the method will be provided in Section 2.2.

The above approach has turned out to be very efficient in a number of important estimates. We will mention four examples in the following.

1. *Doob inequalities.* In [7], Graversen and Peskir used the approach to prove that the inequality

$$||B_\tau^*||_{L^p} \leq \frac{p}{p-1}||B_\tau||_{L^p}, \quad 1 < p < \infty, \tag{2.1.4}$$

holds true for all stopping times $\tau \in L^{p/2}$. Clearly, this estimate is of the form (2.1.2), when both sides are raised to the power p and all the terms are moved on the left-hand side (one has to take $G(x, y) = y^p - \left(\frac{p}{p-1}\right)^p |x|^p$). We refer the reader to the works [12, 14], where the problem is presented from a wider perspective. Consult also the survey [1] by Burkholder, where a related approach is developed.

2. *Hardy-Littlewood inequalities.* For $p = 1$, the inequality (2.1.4) does not hold with any finite constant. Hence it is natural to ask about a substitute for this result in this limiting case. Graversen and Peskir [8] studied the LlogL estimate

$$||B_\tau^*||_{L^1} \leq K\mathbb{E}|B_\tau|\log|B_\tau| + L(K), \quad K > 0,$$

where τ is an arbitrary stopping time belonging to $L^{p/2}$ for some $p > 1$. There are two questions which can be asked:
   1° For which K there is a finite constant $L(K)$ such that the above inequality holds?
   2° For K as in 1°, what is the optimal (i.e., the least) value of $L(K)$?
Graversen and Peskir answered both these questions, by considering $G(x, y) = y - K|x|\log|x| - L(K)$ and showing that the corresponding

gain function U satisfies $U(0,0) \leq 0$ if and only if $K > 1$ and $L(K) \geq K^2/(K-1)e$.

The paper [8] contains the analysis of the related sharp LlogL bound

$$||B_\tau^*||_{L^1} \leq K\mathbb{E}|B_\tau|\log^+|B_\tau| + 1 + \frac{1}{e^K(K-1)}, \quad K > 1,$$

in which the positive part of the logarithm is considered. The proof again exploits the above machinery applied to the gain function G given by $G(x,y) = y - K|x|\log^+|x| - \tilde{L}(K)$. We also refer the interested reader to Peskir's paper [13] which contains the analysis of a related LlogL estimate in which the contribution of B is measured along the whole path (from 0 to τ). See also the work of Gilat [5] devoted to the study of the above LlogL bounds from an analytic point of view.

3. *Mixed norm estimates.* The paper [11] contains further extension of (2.1.4) to the case of different L^p spaces: for any $q > 1$ and any $p \in (0,q)$, Peskir identified the best constant $C_{p,q}$ such that the inequality

$$||B_\tau^*||_{L^q} \leq C_{p,q}||B_\tau||_{L^p}$$

holds for any stopping time $\tau \in L^{p/2}$ (the description of $C_{p,q}$ is a bit complicated, we refer to [11] for the definition). At the first glance the above method does not work here, since the inequality is not of integral form (i.e., it cannot be rewritten as (2.1.2) for any choice of G). To overcome this difficulty, Peskir exploited a homogenization trick. First he studied the estimate

$$\mathbb{E}(B_\tau^*)^q \leq \mathbb{E}|B_\tau|^p + L(p,q),$$

which is of the form (2.1.2) and hence can be handled with the use of the above arguments. Then, using Brownian scaling (for any c, the process $\tilde{B} = cB_{./c^2}$ is a Brownian motion), one gets the estimate

$$\mathbb{E}|B_\tau^*|^q \leq c^{q-p}\mathbb{E}|B_\tau|^p + c^{-p}L(p,q).$$

It remains to optimize over c to get the proof of the $L^p \to L^q$ estimate. See [12, 14] for the presentation of the problem from a wider point of view, consult also the works of Gilat [6] and Jacka [9] which are devoted to the study of the above problem for $p = 1$.

4. The last example presented here is not exactly of the form (2.1.3), but it is closely related to the subject and can serve as a natural object for the further studies, so we have decided to include a brief discussion here. The problem concerns optimal stopping of the *diameter* of the Brownian motion; more precisely, one can study the above moment, logarithmic

and mixed norm inequalities in which the maximum process $(B_t^*)_{t\geq 0}$ is replaced by the "diameter process" $(D_t)_{t\geq 0} = (\sup_{0\leq s\leq t} B_s - \inf_{0\leq s\leq t} B_s)_{t\geq 0}$. To the best of our knowledge, there are two results in this direction. Dubins et al. [4] proved that for any integrable stopping time τ we have

$$||D_\tau||_1 \leq \sqrt{3}||B_\tau||_2$$

and the constant $\sqrt{3}$ is best possible. The corresponding LlogL and weak-type estimates were studied by Osękowski [10]; as the formulation and the constants involved are quite complicated, we refer the interested reader to that paper.

In this chapter we will present a study of related maximal inequalities involving weak Lorentz norms. For any $p \in [1, \infty)$ and any random variable ξ, we define

$$||\xi||_{L^{p,\infty}} = \sup_{\lambda>0} \left[\lambda^p \mathbb{P}(|\xi| \geq \lambda)\right]^{1/p}.$$

These norms (actually, quasinorms) are a little more difficult to handle (from the viewpoint of the above approach), since they are not integral, that is, $||\cdot||_{L^{p,\infty}}$ does not depend directly on any expression of the type $\mathbb{E}\Phi(|\cdot|)$ for some function Φ. However, the situation is easy if one wants to bound weak norms from above: it suffices to control the probabilities $\mathbb{P}(|\cdot| \geq \lambda)$ for each λ. For example, Doob's maximal inequality for martingales (cf. [2]) implies that for any $p \in [1, \infty)$ and any $\tau \in L^{p/2}$ we have

$$||B_\tau^*||_{L^{p,\infty}} \leq ||B_\tau||_{L^p}$$

and the constant 1 is the best. The situation get significantly harder if one wants to provide lower bounds for $||\cdot||_{L^{p,\infty}}$. Our main contribution is to show how to deal with this type of problems and illustrate the approach by proving the following two statements. The first result is a sharp comparison of weak norms.

Theorem 2.1.1. *For any $1 < p < \infty$, any $q \in [1, p]$, and any stopping time $\tau \in L^{p/2}$ we have*

$$||B_\tau^*||_{L^{q,\infty}} \leq \frac{p}{p-1}||B_\tau||_{L^{p,\infty}}. \tag{2.1.5}$$

For any p, q as above, the constant $p/(p-1)$ cannot be improved.

The second statement compares weak and strong norms of stopped B and B^*.

Theorem 2.1.2. *For any* $1 \leq q < p < \infty$ *and any stopping time* τ *we have*

$$||B_\tau^*||_{L^q} \leq \left(\frac{p}{p-q}\right)^{1/q} \frac{p}{p-1}||B_\tau||_{L^{p,\infty}} \qquad (2.1.6)$$

and the constant on the right-hand side is the best possible.

In the next section we present the general method of solving the optimal stopping problems (2.1.3). Section 2.3 contains the description of certain special stopping times which will turn out to be extremal in (2.1.5) and (2.1.6). The remaining two sections are devoted to the proofs of our two main results.

We should emphasize here that although we focus on sharp estimates for Brownian motion, Theorems 2.1.1 and 2.1.2 hold true for general cadlag martingales X and their maximal functions X^* (even without the assumption of the continuity of paths). This follows at once from appropriate embedding theorems (and the fact that if a martingale X is embedded into a Brownian motion B, then the maximal function X^* is majorized by the maximal function B^*). Alternatively, one can check that the reasoning presented in Sections 2.4.2 and 2.5.2 can be extended to the general setting with no difficulty.

2.2 ON OPTIMAL STOPPING OF A PROCESS AND ITS MAXIMAL FUNCTION

In this section we will describe the general approach used in the study of optimal stopping problems for Brownian maximal functions. For a detailed discussion on the subject we refer the interested reader to the monograph [14]. Recall the state space E of the process (B, B^*) given by (2.1.1). Let $G : E \to \mathbb{R}$ be a given gain function and suppose that we are interested in computing the quantity

$$\sup \mathbb{E}G(B_\tau, B_\tau^*), \qquad (2.2.1)$$

where the supremum is taken over all stopping times τ such that the above expectation exists. Observe that $X = (B, B^*)$ is a two-dimensional Markov process which can change (increase) in the second coordinate only after hitting the diagonal $y = |x|$ in E. Off the diagonal, the process X changes only in the first coordinate and thus we may identify it with B. Therefore, the infinitesimal generator of the process can be formally described as follows:

$$\mathbb{L}_X = \frac{\partial^2}{\partial x^2} \quad \text{for } |x| < y,$$

$$\frac{\partial}{\partial y} = 0 \quad \text{for } |x| = y,$$

where $\partial/\partial y$ stands for the right-hand partial derivative with respect to variable y. That is, the generator of X acts on C^2 functions $f : E \to \mathbb{R}$, satisfying

$$\frac{\partial f}{\partial y}(x, |x|) = \lim_{h \downarrow 0} \frac{f(x, |x| + h) - f(x, |x|)}{h} = 0.$$

The first step toward the solution of (2.2.1) is to extend the problem so that the process X can start from arbitrary points from E (this enables the use of Markovian arguments). The extension is straightforward: for any $x \in \mathbb{R}$ and $y \geq |x|$, the process

$$(B_t, B_t^*)^{(x,y)} = \left(x + B_t, y \vee \sup_{0 \leq s \leq t} |x + B_s| \right)$$

starts from the point (x, y), is Markov under $\mathbb{P}$ and hence the class $\mathbb{P}_{x,y} = \text{Law}((B_t, B_t^*)^{(x,y)} | \mathbb{P})$, $(x, y) \in \mathbb{R} \times [0, \infty)$, is a Markovian family of probability measures on the canonical space. Now we extend (2.2.1) and define the associated value function $U : E \to \mathbb{R} \cup \{\infty\}$ by

$$U(x, y) = \sup_\tau \mathbb{E}_{x,y} G(B_\tau, B_\tau^*), \tag{2.2.2}$$

where $\mathbb{E}_{x,y}$ is the expectation with respect to $\mathbb{P}_{x,y}$. Sometimes we will write $\mathbb{E}$ instead of $\mathbb{E}_{0,0}$: the lack of subscripts means that we work in the initial setting in which both B and B^* start from 0. We believe that this will not lead to any confusion.

The reason for extending the original stopping problem (2.2.1) to (2.2.2) is that the resulting value function enjoys certain structural properties which make its explicit identification possible. To describe these properties, introduce the continuation set

$$C = \{(x, y) \in E : U(x, y) > G(x, y)\}$$

and the stopping set

$$D = \{(x, y) \in E : U(x, y) = G(x, y)\}.$$

Clearly, it suffices to find the shape of the continuation region C and the formula for U restricted to this set. To this end, one exploits standard Markovian argumentation and shows that on the continuation set, the value function U must satisfy $\mathbb{L}_X U = 0$. Furthermore, if $(x, |x|) \in C$, then $\frac{\partial U}{\partial y}(x, |x|)$ must be zero. The final comment is that under mild regularity conditions on G, the function U is differentiable at the common boundary of C and D (sometimes in the literature this property is called the principle of smooth fit). These analytic conditions generally determine U only up to one degree of freedom (expressed in terms of the optimal stopping boundary solving a nonlinear differential equation). The maximality principle [12] tells us how to specify the correct U among all the available candidates.

In most examples (see, e.g., the estimates discussed in Section 2.1), the gain function G is of the form $G(x, y) = \Psi(y) - \Phi(|x|)$, where Ψ, $\Phi : [0, \infty) \to \mathbb{R}$ are nondecreasing Borel functions and Φ is convex. In such a case, the solution to the problem (2.2.2) is based on the following two-step procedure.

Step 1. The first step is to use an informal argumentation to get the candidate $\widetilde{U}$ for the value function. Fix a starting point (x, y) and look at the definition of $U(x, y)$:

$$U(x, y) = \sup_{\tau} \left[\mathbb{E}_{x,y} \Psi(B_\tau^*) - \mathbb{E}_{x,y} \Phi(|B_\tau|) \right].$$

By the symmetry of the gain function G, we see that U must also have this property: $U(x, y) = U(-x, y)$ for all $(x, y) \in E$. How can we identify U? Informally speaking we would like to choose τ such that $\mathbb{E}_{x,y} \Psi(B_\tau^*)$ is relatively big, while $\mathbb{E}_{x,y} \Phi(|B_\tau|)$ is relatively small. However, as time increases, so does the quantity $\mathbb{E}_{x,y} \Phi(|B_t|)$, since Φ is convex; in a sense we experience some sort of a "cost of observation". On the other hand, before reaching the "diagonal" $y = |x|$, the process B^* stays at the same level and hence so does $\mathbb{E}_{x,y} \Psi(B_t^*)$. This suggests that we cannot let the process (B, B^*) get too far from the "diagonal" $y = |x|$, since it would be "too expensive" to come back to the diagonal in order to offset the cost spent to travel all that way. In other words, for sufficiently large y there should be a threshold $\gamma(y) \geq 0$ such that if the process (B, B^*) reaches the set $[-\gamma(y), \gamma(y)] \times \{y\}$, we should stop it instantly. In other words, it seems plausible that the stopping region is of the form

$$D = \{(x, y) : |x| \leq \gamma(y), y \geq y_0\}$$

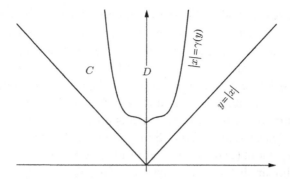

Fig. 2.1 The continuation and stopping regions.

for some number $y_0 \geq 0$ and some function $\gamma : [y_0, \infty) \to \mathbb{R}$ to be found. See Fig. 2.1 which illustrates the geometry of the sets C and D. We have $\tilde{U}(x, y) = \Psi(y) - \Phi(|x|)$ on D (from the very definition of the stopping region), and $\tilde{U}_{xx} = 0$ on C. Hence $\tilde{U}(x, y) = A(y) + B(y)|x|$ on C, for some unknown coefficients $A(\cdot)$, $B(\cdot)$. To identify these, we exploit the principle of smooth fit. If $y \geq y_0$ (that is, $([-y, y] \times \{y\}) \cap D \neq \emptyset$, comparing the left- and right-hand derivatives $U_{x-}(\cdot, y)$, $U_{x+}(\cdot, y)$ at $|x| = \gamma(y)$ implies that

$$\tilde{U}(x, y) = \Psi(y) - \Phi(\gamma(y)) - \Phi'(\gamma(y)+)(|x| - \gamma(y))$$

for $|x| \geq \gamma(y)$. When $y < y_0$, we are forced to take $U(x, y) = U(y_0, y_0)$. Indeed, if we set $\sigma = \inf\{t : B_t^* \geq y_0\}$, then we see that the process $((B_t, B_t^*))_{t \leq \sigma}$ lives in the continuation region and hence, by Itô's formula,

$$U(x, y) = \mathbb{E}_{x,y} U(B_0, B_0^*) = \mathbb{E}_{x,y} U(B_\sigma, B_\sigma^*) = U(y_0, y_0).$$

Thus, we have reduced the problem of identifying U to that of finding the formula for the boundary curve γ. This task is handled with the use of the requirement $\tilde{U}_y(x, |x|) = 0$, which leads to a differential equation for γ. See the monograph [14] for many examples, consult also Section 2.5.

Step 2. The second part of the analysis is to verify rigorously that $\tilde{U}$ has all the necessary properties (which in turn implies that $\tilde{U} = U$). Typically, the argument exploits Itô's formula to prove that for any stopping time τ as in the original problem, the process $Z = \big(\tilde{U}(B_{\tau \wedge t}, B_{\tau \wedge t}^*)\big)_{t \geq 0}$ is a supermartingale. Combining this with the majorization property $\tilde{U} \geq G$, guaranteed by the construction, one proves that

$$\mathbb{E}_{x,y} G(B_\tau, B_\tau^*) \leq \mathbb{E}_{x,y} \tilde{U}(B_\tau, B_\tau^*) \leq \mathbb{E}_{x,y} \tilde{U}(B_0, B_0^*) = \tilde{U}(x, y). \quad (2.2.3)$$

This proves $U \leq \tilde{U}$, and the reverse estimate is usually obtained by considering appropriate example. In many cases, the stopping time $\tau_D = \inf\{t \geq 0 : (B_t, B_t^*) \in D\}$ gives equality in (2.2.3) and hence $U \geq \tilde{U}$, from the very definition of U. Sometimes some minor modifications are necessary to ensure that the stopping time enjoys an appropriate integrability. Since our motivation comes from the estimate (2.1.2), we will mostly be concerned with the case when the pair (B, B^*) starts from the origin $(0, 0)$.

2.3 A FAMILY OF IMPORTANT EXAMPLES

We start with presenting a certain class of stopping times which will serve as extremal examples in the estimates (2.1.5) and (2.1.6). Somewhat similar objects appear in the works of Dubins et al. [4], Dubins and Schwarz [3], Wang [16], and Peskir [13]. Fix constants $a > 0$, $c \in (0, 1)$ and let

$$\tau = \tau_{a,c} = \inf\{t \geq 0 : B_t^* \geq a, |B_t| = cB_t^*\}.$$

We see that the stopping time τ can be split into two parts: first we wait until the maximal function B^* reaches the level a (note that at the time this happens, we have $|B| = B^* = a$) and then we wait until $|B_t|/B_t^*$ drops to the level c. Put $p = (1 - c)^{-1} > 1$. We will prove the following.

Lemma 2.3.1. *For any $r \in (0, p)$ we have $\tau \in L^{r/2}$. Furthermore, we have $\|B_\tau\|_{L^{p,\infty}} = ca$ and for any $\lambda > 0$ the distribution function of $|B_\tau|$ satisfies*

$$\mathbb{P}(|B_\tau| \geq \lambda) = \min\{\|B_\tau\|_{L^{p,\infty}}^p / \lambda^p, 1\}. \tag{2.3.1}$$

Proof. Clearly, we have

$$\mathbb{P}(|B_\tau| \geq \lambda) = 1 \quad \text{if } \lambda \leq ca. \tag{2.3.2}$$

So, suppose that $\lambda > ca$ and consider the function $U : \mathbb{R} \times [0, \infty) \to \mathbb{R}$ given by

$$U(x, y) = \begin{cases} (ca/\lambda)^p & \text{if } y \leq a, \\ p(c/\lambda)^p (x - cy)y^{p-1} & \text{if } a < y \leq \lambda/c, \\ 1 & \text{if } y > \lambda/c. \end{cases}$$

It is easy to see that the pair $(B_{\tau \wedge t}, B_{\tau \wedge t}^*)$ evolves over the set on which $U_{xx} = 0$. Furthermore, we have $U_y(y, y) = 0$ for any $y \geq 0$. Consequently, by Itô's formula, we have

$$\mathbb{E}U(B_{\tau \wedge t}, B_{\tau \wedge t}^*) = U(0, 0) = (ca/\lambda)^p.$$

The function U is bounded and the process $U(B_{\tau \wedge t}, B^*_{\tau \wedge t})$ converges almost surely to $U(B_\tau, B^*_\tau)$, so Lebesgue's dominated convergence theorem yields $\mathbb{E}U(B_\tau, B^*_\tau) = (ca/\lambda)^p$. Finally, we have $U(B_\tau, B^*_\tau) = 1_{\{B^*_\tau \geq \lambda/c\}} = 1_{\{|B_\tau| \geq \lambda\}}$ and hence we obtain

$$\mathbb{P}(|B_\tau| \geq \lambda) = (ca/\lambda)^p \quad \text{if } \lambda > ca.$$

Combining this with (2.3.2), we get that $||B_\tau||_{L^{p,\infty}} = ca$ and the second part of the lemma follows. To deduce the first part, note that we have $B^*_\tau = |B_\tau|/c$ and hence $||B^*_\tau||_{L^{p,\infty}} = ||B_\tau||_{L^{p,\infty}}/c = a$. However, for any $r < p$ we have $L^r \subset L^{p,\infty}$; thus $||B^*_\tau||_{L^r} < \infty$ and the assertion follows from Burkholder-Davis-Gundy inequality (cf. [15]). $\qquad\square$

2.4 PROOF OF THEOREM 2.1.1

We are ready to establish the first of the two theorems announced in Section 2.1. Fix $p \in (1, \infty)$. Clearly, when proving (2.1.5), it is enough to consider $q = p$, since $||B^*_\tau||_{L^{q,\infty}} \leq ||B^*_\tau||_{L^{p,\infty}}$. We want to show that the least constant C_p, for which

$$\sup_{\tau \in L^{p/2}} \left[||B^*_\tau||_{L^{p,\infty}} - C_p ||B_\tau||_{L^{p,\infty}} \right] \leq 0 \qquad (2.4.1)$$

is equal to $p/(p-1)$. Clearly, no manipulations on the left-hand side of (2.4.1) transform it into the form (2.1.3), so the general methodology does not work. It is easy to get rid of the weak norm of the variable B^*_τ. Namely, by homogeneity and Brownian scaling, it is enough to establish the inequality

$$\sup_{\tau \in L^{p/2}} \left[\mathbb{P}(B^*_\tau \geq 1) - C_p^p ||B_\tau||^p_{L^{p,\infty}} \right] \leq 0. \qquad (2.4.2)$$

However, due to the appearance of the term $||B_\tau||_{L^{p,\infty}}$, the problem persists. The idea is to replace the troublesome term with a slightly smaller quantity so that the expression gets the appropriate form and such that the extremal stopping times in both cases will be the same. To be more precise, fix a continuous, nondecreasing function $\Phi : [0, \infty) \to [0, \infty)$ and consider the inequality

$$\sup_{\tau \in L^{p/2}} \left[\mathbb{P}(B^*_\tau \geq 1) - \mathbb{E}\Phi(|B_\tau|) \right] \leq 0. \qquad (2.4.3)$$

The problem (2.4.3) can be studied with the use of optimal stopping techniques, since it corresponds to the choice $G(x, y) = 1_{\{y \geq 1\}} - \Phi(|x|)$.

Suppose we have successfully established (2.4.3) and assume that for any nonnegative random variable X we have the inequality

$$\mathbb{E}\Phi(X) \le C_p^p \|X\|_{L^{p,\infty}}^p. \qquad (2.4.4)$$

Then the estimate (2.4.2) will follow. Of course, it is absolutely not clear whether such a function Φ exists and how to search for it. A crucial (but a little imprecise) observation is that if we search for best constant C_p, then there must be an optimal (or almost optimal) nontrivial stopping time τ which will give equality in (2.4.1); if Φ has all the required properties, then τ must necessarily be (almost) optimal in (2.4.3) and must also return equality in (2.4.4) (with $X = |B_\tau|$).

We split the remainder of this section into two parts. The first part is informal and serves as a description of intuitive arguments which lead to the discovery of the special function Φ. The second part is devoted to a formal proof of Theorem 2.1.1.

2.4.1 On the Search for Φ

We should stress here that the reasoning we are going to present will be informal: the arguments leading to the formula for Φ will be based on a number of assumptions and guesses (e.g., we will treat the value function of the optimal stopping problem (2.4.3) as a smooth function). The formal verification of all the properties will be postponed to the next subsection.

A lot of information about Φ is contained in the estimate (2.4.4). We can rewrite this inequality in the form

$$\int_0^\infty \Phi'(\lambda)\mathbb{P}(X \ge \lambda)\mathrm{d}\lambda \le C_p^p \|X\|_{L^{p,\infty}}^p.$$

It is clear that if we fix $\|X\|_{L^{p,\infty}}$, then the left-hand side is maximized by a random variable with the distribution satisfying $\mathbb{P}(X \ge \lambda) = \min\{\|X\|_{L^{p,\infty}}^p \lambda^{-p}, 1\}$. This strongly suggests that the optimal stopping time τ in (2.4.1) should have the property that $|B_\tau|$ has this distribution. In the light of the examples analyzed in the preceding section, it is natural to conjecture that the stopping region for the problem (2.4.3) is

$$D = \{(x,y) \in E : y \ge a, x \le cy\},$$

where $c = 1 - p^{-1}$ and $a > 0$ is a fixed constant. Combining this information with the discussion in Section 2.2, we see that the value function of the problem (2.4.3) should be equal to

$$U(x, y) = \begin{cases} 1_{\{y \geq 1\}} - \Phi(|x|) & \text{if } y \geq a, x \leq cy, \\ 1_{\{y \geq 1\}} - \Phi(cy) - \Phi'(cy)(|x| - cy) & \text{if } y \geq a, x \geq cy, \\ 1_{\{a \geq 1\}} - \Phi(ca) - \Phi'(ca)(a - ca) & \text{if } y < a. \end{cases}$$

It seems reasonable to assume that $a = 1$; otherwise there might be problems with ensuring the continuity of U. Under this condition, we have

$$U(x, y) = \begin{cases} 1 - \Phi(|x|) & \text{if } y \geq 1, x \leq cy, \\ 1 - \Phi(cy) - \Phi'(cy)(|x| - cy) & \text{if } y \geq 1, x \geq cy, \quad (2.4.5) \\ 1 - \Phi(c) - \Phi'(c)(1 - c) & \text{if } y < 1. \end{cases}$$

Now, the equality $U_y(y, y) = 0$ for $y \geq 1$ implies $\Phi''(cy) = 0$ and hence Φ is linear on $[c, \infty)$: $\Phi(y) = \alpha y + \beta$ there. What about the interval $[0, c)$? A little thought and experimentation suggests that we should take $\Phi(c) = 0$ and extend Φ to the whole $[0, \infty)$ by setting $\Phi \equiv 0$ on $[0, c]$. This implies $\alpha c + \beta = 0$, and the inequality $U(0, 0) \leq 0$ (which we want to have at the very end), implies $\alpha \geq p$. We assume that actually equality holds here: $\alpha = p$, and this gives us the final formula for Φ: $\Phi(y) = p(y - 1 + p^{-1})_+$.

2.4.2 Proof of Theorem 2.1.1

Now we are ready for the formal proof of the weak-type estimate. Let U be the special function constructed in the preceding subsection. Directly from the above analysis, we have $U_y(y, y) = 0$ for any $y \geq 0$. Furthermore, it is easy to see that for a fixed y, the function $x \mapsto U(x, y)$ is concave on $[-y, y]$. Therefore, by Itô's formula, for any stopping time $\tau \in L^{p/2}$ and any $t \geq 0$ we have

$$\mathbb{E}U(B_{\tau \wedge t}, B^*_{\tau \wedge t}) \leq U(0, 0) = 0.$$

Clearly, we have $U(x, y) \leq A(1 + |x| + y)$ for some universal constant A; furthermore, the condition $\tau \in L^{p/2}$ implies that $B^*_\tau \in L^p$, by Burkholder-Davis-Gundy inequality. Consequently, if we let $t \to \infty$ above, Lebesgue's dominated convergence theorem yields

$$\mathbb{E}U(B_\tau, B^*_\tau) \leq 0. \quad (2.4.6)$$

On the other hand, we have the majorization

$$U(x, y) \geq 1_{\{y \geq 1\}} - \Phi(|x|), \quad |x| \leq y. \quad (2.4.7)$$

Actually, for $y \geq 1$ both sides are equal; if $y < 1$, the inequality is equivalent to $\Phi(|x|) \geq 0$, which is evident. Combining (2.4.7) with (2.4.6) yields

$$\mathbb{P}(B_\tau^* \geq 1) - \mathbb{E}\Phi(|B_\tau|) \leq 0.$$

This implies

$$\mathbb{E}\Phi(|B_\tau|) = \int_0^\infty \Phi'(\lambda)\mathbb{P}(|B_\tau| \geq \lambda)\mathrm{d}\lambda$$

$$\leq p \int_{1-p^{-1}}^\infty \frac{||B_\tau||_{L^{p,\infty}}^p}{\lambda^p}\mathrm{d}\lambda = \left(\frac{p}{p-1}\right)^p ||B_\tau||_{L^{p,\infty}}^p$$

and hence we get (2.4.2), with $C_p = p/(p-1)$.

It remains to verify that the constant $p/(p-1)$ cannot be improved in (2.1.5) (for general p, q as in the statement of the theorem). This time it suffices to deal with the case $q = 1$. Consider the example from Section 2.3, with $c = 1 - p^{-1}$ and $a = 1$. We have

$$||B_\tau^*||_{L^{1,\infty}} \geq \mathbb{P}(B_\tau^* \geq 1) = 1 = \frac{p}{p-1}||B_\tau||_{L^{p,\infty}},$$

which completes the proof.

2.5 PROOF OF THEOREM 2.1.2

As in the preceding section, it is convenient to split the reasoning into two parts. In the first part we construct the appropriate auxiliary optimal stopping problem and present the informal search for its solution. The second part is devoted to the rigorous proof of the estimate (2.1.6).

2.5.1 An Auxiliary Optimal Stopping Problem

Due to certain technical reasons, throughout this subsection we assume that $q > 1$; the case $q = 1$ will be obtained via standard limiting arguments. Clearly, in the study of (2.1.6), the method described in Section 2.2 cannot be applied directly. The first problem is that both sides of (2.1.6) involve norms with different exponents p and q; the second issue is the appearance of the weak norms. To deal with the first difficulty, fix $K > 0$ and consider the problem

$$\sup_\tau \left[||B_\tau^*||_{L^q}^q - K||B_\tau||_{L^{p,\infty}}^p \right], \qquad (2.5.1)$$

where the supremum is taken over all stopping times $\tau \in L^{p/2}$. We expect that after an appropriate choice of the constant K, the solution to (2.5.1) will establish the original statement. To handle the weak norms, we consider the auxiliary stopping problem

$$\sup_{\tau} \mathbb{E}\big[(B_\tau^*)^q - K\Phi(|B_\tau|)\big], \qquad (2.5.2)$$

where the supremum is taken over the same class of stopping times as previously, and $\Phi : [0,\infty) \to [0,\infty)$ is a C^1 nondecreasing convex function which will be specified later. Obviously, this problem can be studied with the use of the general methodology: suppose temporarily that we have successfully solved it and let us proceed with the analysis of (2.5.1). If Φ is chosen so that

$$\mathbb{E}\Phi(X) \leq ||X||_{L^{p,\infty}}^p \qquad (2.5.3)$$

for any nonnegative random variable X, then

$$\sup_{\tau} \left[||B_\tau^*||_{L^q}^q - K||B_\tau||_{L^{p,\infty}}^p \right] \leq \sup_{\tau} \mathbb{E}\big[(B_\tau^*)^q - K\Phi(|B_\tau|)\big]. \quad (2.5.4)$$

Now, we make analogous remark to that from the preceding section. If there is a stopping time τ for which the supremum in (2.5.2) is attained and for which we have $\mathbb{E}\Phi(|B_\tau|) = ||B_\tau||_{L^{p,\infty}}^p$, then we have equality in (2.5.4) and hence we should be done.

The above analysis reduces the problem to that of finding an appropriate function Φ. The key information on this object is contained in the inequality (2.5.3). If Φ is of class C^1 and nondecreasing, as we have assumed above, we may rewrite this inequality in the form

$$\int_0^\infty \Phi'(\lambda)\mathbb{P}(X \geq \lambda)d\lambda \leq ||X||_{L^{p,\infty}}^p.$$

Clearly, the left-hand side is maximal when we take the variable X with the distribution $\mathbb{P}(X \geq \lambda) = \min\{||X||_{L^{p,\infty}}^p \lambda^{-p}, 1\}$. So, the extremal stopping time τ should be chosen so that $|B_\tau|$ has this distribution. In the light of the examples presented in Section 2.3, we get that the stopping region of the problem

$$U(x,y) = \sup_{\tau \in L^{p/2}} \mathbb{E}_{x,y}\big[(B_\tau^*)^q - K\Phi(|B_\tau|)\big] \qquad (2.5.5)$$

should be equal to

$$D = \big\{(x,y) \in \mathbb{R} \times [0,\infty) : y \geq a \text{ and } |x| \leq cy\big\},$$

where $c = 1 - p^{-1}$ and a is a certain positive constant. Now recall the discussion presented at the end of Section 2.2. It implies that U should be given by the formula

$$U(x,y) = \begin{cases} y^q - K\Phi(|x|) & \text{if } y \geq a, |x| \leq cy, \\ y^q - K\Phi(cy) - K\Phi'(cy)(|x| - cy) & \text{if } y \geq a, |x| \geq cy, \\ a^q - K\Phi(ca) - K\Phi'(ca)(a - ca) & \text{if } y < a. \end{cases}$$

The equality $U_y(y,y) = 0$ is equivalent to saying that $K\Phi''(cy) = \frac{q}{c(1-c)}y^{q-2}$ for $y \geq a$. Solving this differential equation, we get that if $y \geq ca$, then

$$K\Phi(y) = \frac{y^q}{c^{q-1}(1-c)(q-1)} + \alpha y + \beta,$$

for some constants α, β at our disposal. We take them so that $\Phi(ca) = \Phi'(ca) = 0$, and obtain

$$K\Phi(y) = \frac{y^q - q(ca)^{q-1}y + (q-1)(ca)^q}{c^{q-1}(1-c)(q-1)}.$$

We extend Φ to the whole $[0, \infty)$ by setting $\Phi \equiv 0$ on $[0, ca]$. Thus, finally, we get that

$$\Phi(y) = \left(\frac{y^q - q(ca)^{q-1}y + (q-1)(ca)^q}{Kc^{q-1}(1-c)(q-1)} \right)_+.$$

To get the value of the constant K, we return to the inequality (2.5.3) and recall that the random variable $|B_\tau|$, with τ being the extremal stopping time, should give equality there. As we have computed in Section 2.3, the stopping time τ corresponding to the above choice of D satisfies $\|B_\tau\|_{L^{p,\infty}} = ca$, so we get

$$(ca)^q = \int_{ca}^{\infty} \Phi'(\lambda) \cdot \frac{(ca)^p}{\lambda^p} d\lambda = \frac{\frac{q}{p-q}(ca)^q - \frac{q}{p-1}(ca)^q}{Kc^{q-1}(1-c)(q-1)}.$$

Hence, recalling that $c = 1 - p^{-1}$, after some straightforward computations we obtain

$$K = \frac{q}{p-q} \left(\frac{p}{p-1} \right)^q.$$

Plugging all this information into the definition of U, we see that this function is given by the formula

$$U(x, y)$$
$$= \begin{cases} y^q - \frac{y^q - q(ca)^{q-1}y + (q-1)(ca)^q}{c^{q-1}(1-c)(q-1)} & \text{if } y \geq a, |x| \leq cy, \\ py^q + pc^{p+1-q}a^p - \frac{pq}{q-1}|x|y^{q-1} + pqa^{q-1}|x| & \text{if } y \geq a, |x| \geq cy, \\ a^q & \text{if } y < a. \end{cases}$$

This time we do not specify the value of the parameter a. In contrast with the weak type estimate from the previous section, we need the whole family of auxiliary optimal stopping problems (2.5.5) (corresponding to different choices of a) to establish the desired inequality.

2.5.2 Formal Proof of (2.1.6)

Suppose first that $q > 1$. Take an arbitrary stopping time $\tau \in L^{p/2}$ and let $a = ||B_\tau||_{L^{p,\infty}} \cdot \frac{p}{p-1}$. Let U be the function constructed in the previous section. One easily checks that U is of class C^1, for each y the function $x \mapsto U(x, y)$ is concave on $[-y, y]$ and we have $U_y(y, y) = 0$. Consequently, Itô formula yields

$$\mathbb{E}U(B_{\tau \wedge t}, B^*_{\tau \wedge t}) \leq U(0, 0) = a^q$$

for any $t \geq 0$. We have $U(x, y) \leq A(1 + |x|^q + y^q)$ for some universal constant A (depending only on p, q, and a) and, by Burkholder-Davis-Gundy inequality, we get $B^*_\tau \in L^p \subset L^q$. Consequently, letting $t \to \infty$ above, we obtain

$$\mathbb{E}U(B_\tau, B^*_\tau) \leq a^q \qquad (2.5.6)$$

in the light of Lebesgue's dominated convergence theorem. On the other hand, we have $U(x, y) \geq y^q - K\Phi(|x|)$ on E. This is clear from the very construction when $y \geq a$, for remaining (x, y) the majorization is equivalent to

$$a^q \geq y^q - \Phi(|x|),$$

which is evident: Φ is nonnegative and $y < a$. Therefore, (2.5.6) and (2.5.3) yield

$$\mathbb{E}(B^*_\tau)^q \leq K\Phi(|B_\tau|) + a^q \leq K(ca)^q + a^q$$
$$= \frac{p}{p-q}a^q = \frac{p}{p-q}\left(\frac{p}{p-1}\right)^q ||X||^q_{L^{p,\infty}}.$$

This completes the proof of (2.1.6) in the case $q > 1$. Letting $q \downarrow 1$ we immediately get the assertion in the limit case $L^{p,\infty} \to L^1$. To show that the inequality is sharp, we exploit the exemplary τ from Section 2.3 with $a = 1$

and $c = 1 - p^{-1}$. Directly from the construction we see that $B_\tau^* = |B_\tau|/c$ and hence, by Lemma 2.3.1,

$$||B_\tau^*||_{L^q} = c^{-1}||B_\tau||_{L^q} = c^{-1}\left(q\int_0^\infty \lambda^{q-1}\mathbb{P}(|B_\tau| \geq \lambda)d\lambda\right)^{1/q}$$

$$= \left(\frac{p}{p-q}\right)^{1/q} = \left(\frac{p}{p-q}\right)^{1/q}\frac{p}{p-1}||B_\tau||_{L^{p,\infty}}.$$

This completes the proof of Theorem 2.1.2.

REFERENCES

[1] D.L. Burkholder, Explorations in martingale theory and its applications, in: Ecole d'Eté de Probabilités de Saint Flour XIX-1989, Lecture Notes in Mathematics, vol. 1464, 1991, pp. 1–66.

[2] J.L. Doob, Stochastic Processes, Wiley, New York, 1953.

[3] L.E. Dubins, G. Schwarz, A sharp inequality for sub-martingales and stopping times, Soc. Math. de France Astérisque 157/158 (1988) 129–145.

[4] L.E. Dubins, D. Gilat, I. Meilijson, On the expected diameter of an L_2-bounded martingale, Ann. Probab. 37 (1) (2009) 393–402.

[5] D. Gilat, The best bound in the LlogL inequality of Hardy and Littlewood and its martingale counterpart, Proc. Am. Math. Soc. 97 (3) (1986) 429–436.

[6] D. Gilat, On the ratio of the expected maximum of a martingale and the L_p-norm of its last term, Israel J. Math. 63 (3) (1988) 270–280.

[7] S.E. Graversen, G. Peskir, On Doob's maximal inequality for Brownian motion, Stoch. Process. Appl. 69 (1) (1997) 111–125.

[8] S.E. Graversen, G. Peskir, Optimal stopping in the LlogL-inequality of Hardy and Littlewood, Bull. Lond. Math. Soc. 30 (2) (1998) 171–181.

[9] S.D. Jacka, Optimal stopping and best constants for Doob-like inequalities. I. The case $p = 1$, Ann. Probab. 19 (4) (1991) 1798–1821.

[10] A. Osękowski, Estimates for the diameter of a martingale, Stochastics 87 (2015) 235–256.

[11] G. Peskir, The best Doob-type bounds for the maximum of Brownian paths, in: High Dimensional Probability (Oberwolfach, 1996), Progr. Probab., vol. 43, Birkhäuser, Basel, 1998, pp. 287–296.

[12] G. Peskir, Optimal stopping of the maximum process: the maximality principle, Ann. Probab. 26 (4) (1998) 1614–1640.

[13] G. Peskir, The integral analogue of the Hardy-Littlewood LlogL-inequality for Brownian motion, Math. Inequal. Appl. 1 (1) (1998) 137–148.

[14] G. Peskir, A. Shiryaev, Optimal stopping and free-boundary problems, in: Lectures in Mathematics ETH Zürich, Birkhäuser Verlag, Basel, 2006.

[15] D. Revuz, M. Yor, Continuous Martingales and Brownian Motion, third ed., Springer, Berlin, 1999.

[16] G. Wang, Sharp maximal inequalities for conditionally symmetric martingales and Brownian motion, Proc. Am. Math. Soc. 112 (1991) 579–586.

CHAPTER 3

On the Absolute Constants in Nagaev–Bikelis-Type Inequalities

Irina Shevtsova[*,†,‡]

[*]Hangzhou Dianzi University, Hangzhou, China
[†]Moscow State University, Moscow, Russia
[‡]Institute for Informatics Problems FRC IC RAS, Moscow, Russia

3.1 INTRODUCTION AND FORMULATION OF THE MAIN RESULTS

3.1.1 Notation

For $0 \leqslant \delta \leqslant 1$ by $\mathcal{F}_{2+\delta}$ denote the set of all distribution functions (d.f.'s) on $\mathbb{R}$ satisfying

$$\int x \mathrm{d}F(x) = 0, \quad \int |x|^{2+\delta} \mathrm{d}F(x) < \infty.$$

Let $X_1, X_2, \ldots, X_n$ be independent random variables (r.v.'s) with d.f.'s $F_1, F_2, \ldots, F_n \in \mathcal{F}_{2+\delta}$ for some $\delta \in [0, 1]$,

$$\sigma_j^2 = \mathsf{E}X_j^2, \quad \beta_{2+\delta, j} = \mathsf{E}|X_j|^{2+\delta}, \quad j = 1, 2, \ldots, n, \quad B_n^2 = \sum_{j=1}^{n} \sigma_j^2 > 0,$$

$$\ell_n = \sum_{j=1}^{n} \frac{\beta_{2+\delta, j}}{B_n^{2+\delta}}, \quad \tau_n = \sum_{j=1}^{n} \frac{\sigma_j^{2+\delta}}{B_n^{2+\delta}}, \quad L_n(z) = \sum_{j=1}^{n} \mathsf{E}X_j^2 \mathbf{1}(|X_j| > z), \ z \geqslant 0,$$

$$\varphi(x) = \frac{1}{\sqrt{2\pi}} e^{-x^2/2}, \quad \Phi(x) = \int_{-\infty}^{x} \varphi(t)\mathrm{d}t,$$

$$\overline{F}_n(x) = \mathsf{P}(X_1 + \cdots + X_n < x B_n),$$

$$\Delta_n(x) = |\overline{F}_n(x) - \Phi(x)|, \quad \Delta_n = \sup_{x \in \mathbb{R}} \Delta_n(x), \quad n \in \mathbb{N}.$$

Note that $\ell_n = \tau_n = 1$ for $\delta = 0$. The quantities ℓ_n and $L_n(z)$ are called the *Lyapunov* and the *Lindeberg* fractions, respectively. It is easy to see that $\ell_n \geqslant \tau_n$ and

Inequalities and Extremal Problems in Probability and Statistics. http://dx.doi.org/10.1016/B978-0-12-809818-9.00003-3

$$\frac{1}{B_n}\int_0^{B_n} L_n(z)\mathrm{d}z = \sum_{j=1}^n \left[\mathsf{E}X_j^2\mathbf{1}(|X_j| > B_n) + \frac{\mathsf{E}|X_j|^3\mathbf{1}(|X_j| \leqslant B_n)}{B_n}\right]$$

$$\leqslant \sum_{j=1}^n \frac{\beta_{2+\delta,\,j}}{B_n^\delta} = B_n^2 \ell_n$$

for every $\delta \in [0, 1]$.

3.1.2 A Short Historical Review of the Uniform Estimates

In the present subsection a short review of the estimates for the Kolmogorov distance Δ_n are given, which will also have its uses in the next sections.

First of all one should cite a celebrated Berry–Esseen inequality (see, e.g., [55]) which states that under the above conditions, for every $\delta \in [0, 1]$ there exist absolute constants $C_0(\delta)$ such that

$$\Delta_n \leqslant C_0(\delta)\ell_n \quad \text{for every } n \in \mathbb{N} \text{ and } F_1, \dots, F_n \in \mathcal{F}_{2+\delta}. \tag{3.1.1}$$

Inequality (3.1.1) was proved for the first time with $\delta = 1$ independently by Berry [3] (in the i.i.d. case) and Esseen [11] (in the general situation). For $\delta = 0$ inequality (3.1.1) is trivial, since in this case $\ell_n = 1$ and Δ_n is always bounded. For $0 < \delta < 1$ inequality (3.1.1) may be deduced from the estimate

$$\Delta_n \leqslant A_1(\delta)\left(\ell_n + (\ell_n)^{1/\delta}\right),$$

obtained by Esseen [12], where $A_1(\delta)$ depends only on δ. Inequality (3.1.1) may also be deduced from the estimate

$$\Delta_n \leqslant \frac{A_2}{B_n^2 g(B_n)} \sum_{k=1}^n \mathsf{E}X_k^2 g(X_k), \quad n \in \mathbb{N}, \ g \in \mathcal{G} \tag{3.1.2}$$

with $g(x) = |x|^\delta$, which was proved by Katz [19] in 1963 in the i.i.d. case and generalized by Petrov [54] in 1965 to the non-i.i.d. case, where A_2 is an absolute constant and $\mathcal{G}$ is a set of all even functions $g \colon \mathbb{R} \to \mathbb{R}_+$ such that both $g(x)$ and $x/g(x)$ do not decrease on $\mathbb{R}_+$.

In 1966 Osipov [43] (see also [2, 8, 14, 49–51, 57]) proved an estimate

$$\Delta_n \leqslant A_3 \inf_{\varepsilon > 0} \sum_{k=1}^n \left[\frac{\mathsf{E}X_k^2\mathbf{1}(|X_k| > \varepsilon B_n)}{B_n^2} + \frac{\mathsf{E}|X_k|^3\mathbf{1}(|X_k| \leqslant \varepsilon B_n)}{B_n^3}\right], \tag{3.1.3}$$

for all $F_1, \dots, F_n \in \mathcal{F}_2$ and $n \in \mathbb{N}$, where A_3 is some absolute constant. The proof of (3.1.3) given in [43] was self-contained and independent of

Katz–Petrov's inequality (3.1.2). However, as it can easily be made sure, the greatest lower bound on the right-hand side of (3.1.3) is attained at $\varepsilon = 1$, more precisely,

$$\mathsf{E}X^2\mathbf{1}(|X| > 1) + \mathsf{E}|X|^3\mathbf{1}(|X| \leqslant 1) = \mathsf{E}X^2\min\{|X|, 1\}(\mathbf{1}(X \in B)$$
$$+ \mathbf{1}(X \notin B)) \leqslant \mathsf{E}X^2\mathbf{1}(X \in B) + \mathsf{E}|X|^3\mathbf{1}(X \notin B)$$

for arbitrary measurable $B \subset \mathbb{R}$ and every r.v. X with $\mathsf{E}X^2 < \infty$ (apparently, this fact was noticed for the first time by Loh [27]), so that inequality (3.1.3) is a corollary to the estimate

$$\Delta_n \leqslant A_3 \sum_{k=1}^{n} \left[\frac{\mathsf{E}X_k^2\mathbf{1}(|X_k| > B_n)}{B_n^2} + \frac{\mathsf{E}|X_k|^3\mathbf{1}(|X_k| \leqslant B_n)}{B_n^3} \right] = \frac{1}{B_n^3}\int_0^{B_n} L_n(z)\mathrm{d}z,$$

$$(3.1.4)$$

which, in its turn, trivially follows from Katz–Petrov inequality (3.1.2) with $g(x) = \min\{|x|/B_n, 1\} \in \mathcal{G}$. Applying the same reasoning as in [50, 56], Korolev and Popov [24] proved the function $g(x) = \min\{|x|/B_n, 1\}$ to be extremal in (3.1.2) (i.e., minimizing the right-hand side of (3.1.2)), yielding $A_2 = A_3$, and established an upper bound $A_3 \leqslant 2.011$, improving the previous estimates in [2, 8, 14, 49–51]. In 2015, this bound was slightly improved by Dorofeyeva and Korolev [22] to $A_3 \leqslant 1.8627$. Since $C_0(\delta) \leqslant A_3$, inequality (3.1.3) (or (3.1.2)) yields a uniform upper bound for $C_0(\delta) \leqslant 1.8627$ which holds for all $\delta \in (0, 1]$.

Inequality (3.1.3) can be regarded as a *natural* convergence rate estimate in the Lindeberg–Feller theorem, since for uniformly asymptotically negligible random summands, satisfying the Feller condition $\lim_{n\to\infty} \max_{1\leqslant k\leqslant n} \sigma_k^2/B_n^2 = 0$, the left- and right-hand sides of (3.1.3) either are both infinitesimal or none of them tends to zero. Indeed, the right-hand side of (3.1.3) does not exceed $A_3(\varepsilon + B_n^{-2}L_n(\varepsilon B_n))$ which can be made arbitrarily small iff the Lindeberg condition, $\sup_{z>0} \lim_{n\to\infty} L_n(z) = 0$, holds. On the other hand, the Lindeberg condition is equivalent to that the CLT and the Feller condition hold.

There are a lot of works devoted to the estimation of the constant $C_0(1)$ in the classical Berry–Esseen inequality (3.1.1) (e.g., see the historical reviews in [21, 25, 69]). During the last decade, by the efforts of Shevtsova, Korolev [21, 25, 26, 65, 67, 68, 70], and Tyurin [75–79] the upper bound for $C_0(1)$ was lowered to 0.469 in the i.i.d. case and to 0.5583 in the general case [70]. Moreover, in [70] there were evaluated the constants in the so-called Berry–Esseen-type inequalities with an improved structure in the form

$$\Delta_n \leqslant \begin{cases} \min\{0.3723(\ell_n + 0.5\tau_n), 0.3057(\ell_n + \tau_n)\}, & F_1, \ldots, F_n \in \mathcal{F}_3, \\ \min\{0.3322(\ell_n + 0.429\tau_n), 0.3031(\ell_n + 0.646\tau_n)\}, & F_1 = \cdots = F_n \in \mathcal{F}_3, \end{cases}$$

$$(3.1.5)$$

improving the pioneer results in [21, 25, 26]. It is interesting to note that the factors of the Lyapunov fraction ℓ_n in (3.1.5) are strictly less than the lower bound

$$C_0(1) \geqslant \frac{\sqrt{10}+3}{6\sqrt{2\pi}} = 0.4097\ldots$$

for the absolute constant $C_0(1)$ in the classical Berry–Esseen inequality, discovered by Esseen [13].

In [71] there were proposed generalizations of (3.1.5) to the case of arbitrary $\delta \in (0, 1]$ in the form

$$\Delta_n \leqslant \inf_{s \in \mathbb{R}_+} C_s(\delta) \cdot (\ell_n + s\tau_n), \quad F_1, \ldots, F_n \in \mathcal{F}_{2+\delta}, \ n \in \mathbb{N}, \quad (3.1.6)$$

yielding the upper bounds for $C_0(\delta) \leqslant \inf_{s \geqslant 0}(1+s)C_s(\delta)$ due to $\ell_n \leqslant \tau_n$. These bounds remain the best known up till now. The infimums over $s \in \mathbb{R}_+$ in (3.1.6) are attained at some $s = s_0(\delta) \in [0, 1]$ within the framework of the numerical method used in [71] (see Table 3.2). Moreover, in [71] there were also computed the values of $s = s_1(\delta)$ that minimize the constants $C_s(\delta)$ themselves, still within the framework of the method used (see Table 3.2). The lower bounds for the constants $C_s(\delta)$ which hold true even in the asymptotic sense were discovered in [69, 71]. For example, in [71] in terms of the so-called *conditional upper asymptotically exact constant* it was proved that

$$\inf_{s \geqslant 0} C_s(\delta) \geqslant \sup_{\gamma > 0, \, m \in \mathbb{N} \cup \{0\}} \gamma^{\delta/2}\left(e^{-\gamma}\sum_{k=0}^{m}\frac{\gamma^k}{k!} - \Phi\left(\frac{m-\gamma}{\sqrt{\gamma}}\right)\right), \quad 0 \leqslant \delta \leqslant 1.$$

$$(3.1.7)$$

In particular, for $\delta = 1$ with $m = 6$ and $\gamma = 6.4206\ldots$ we have

$$\inf_{s \geqslant 0} C_s(1) \geqslant 0.266012\ldots = \frac{2}{3\sqrt{2\pi}} + 0.0000505\ldots$$

The values of the lower bounds in (3.1.7) for some other values of δ are given in the second column of Table 3.1.

A more detailed historical review of the uniform estimates can be found in [21, 23, 25, 66].

Table 3.1 Lower Bounds for $\inf_{s \geqslant 0} C_s(\delta)$ Given in (3.1.7) (Second Column) and Upper Bounds for the Constants $C_s(\delta)$ From (3.1.6) for Some $s \in [0, 1]$ and $\delta \in (0, 1)$ Originally Obtained in [70, 71]

δ	$C_s(\delta) \geqslant$	Upper Bounds (i.i.d. Case)				Upper Bounds (Non-i.i.d. Case)			
		s_0	$C_{s_0}(\delta)$	s_1	$C_{s_1}(\delta)$	s_0	$C_{s_0}(\delta)$	s_1	$C_{s_1}(\delta)$
1	0.26601	0	0.4690	0.646	0.3031	0	0.5583	1	0.3057
0.9	0.2698	0.410	0.3514	0.626	0.3073	0.52	0.37046	1	0.3108
0.8	0.2819	0.6356	0.3166	0.6356	0.3166	0.15	0.49939	1	0.3215
0.7	0.2961	0.5830	0.3306	0.5830	0.3306	0.06	0.5572	1	0.3367
0.6	0.3128	0.5131	0.3492	0.5131	0.3492	0.02	0.60313	0.859	0.3557
0.5	0.3328	0.4444	0.3728	0.4444	0.3728	0.01	0.6432	0.834	0.3795
0.4	0.3568	0.3652	0.4025	0.47	0.4022	0.02	0.6828	0.806	0.4091
0.3	0.3862	0.2823	0.4399	0.52	0.4384	0.04	0.7229	0.778	0.4457
0.2	0.4232	0.1920	0.4868	0.58	0.4828	0.06	0.7666	0.748	0.4905
0.1	0.4714	0.0744	0.5439	0.63	0.5372	0.08	0.81388	0.710	0.5454

Table 3.2 Values of $(1 + s_0(\delta))C_{s_0}(\delta)$ From [70, 71] Rounded Up, Which Serve as the Best Known Upper Bounds for the Constants $C_0(\delta)$ in (3.1.6)

δ	I.i.d. $C_0(\delta) \leqslant$	Non-i.i.d. $C_0(\delta) \leqslant$	δ	I.i.d. $C_0(\delta) \leqslant$	Non-i.i.d. $C_0(\delta) \leqslant$
1	0.4690	0.5583	0.5	0.5385	0.6497
1−	0.4748	0.5591	0.4	0.5495	0.6965
0.9	0.4955	0.5631	0.3	0.5641	0.7519
0.8	0.5179	0.5743	0.2	0.5798	0.8126
0.7	0.5234	0.5907	0.1	0.5842	0.8790
0.6	0.5284	0.6152	0	0.5410	0.5410

3.1.3 Nonuniform Estimates: History, Problem Statements, Formulation of the New Results

Investigation of the dependence of the remainder term $\Delta_n(x)$ on n and x started as far back as by Cramér [10] for i.i.d. random summands with exponentially decreasing tails. For distributions satisfying the power-type moment conditions considered in the present study, historically the first estimate for $\Delta_n(x)$, apparently, was obtained by Esseen [12] in the i.i.d. case with $\delta = 1$ in the form

$$\Delta_n(x) \leqslant A_4 \left(\frac{\beta_{3,1}}{\sigma_1^3} \right) \cdot \frac{\ln(2 + |x|)}{(1 + |x|^3)\sqrt{n}}, \quad x \in \mathbb{R},$$

where $A_4(\cdot)$ depends only on the argument inside the brackets. By use of a new smoothing inequality, Meshalkin and Rogozin [29] proved existence of absolute constants A_5 and A_6 such that for all $n \geqslant 1$ and $F_1 = \cdots = F_n \in \mathcal{F}_3$

$$\Delta_n(x) \leqslant A_5 \cdot \frac{\beta_{3,1}}{\sigma_1^3 \sqrt{n}} \cdot \frac{\max\{\ln n, \ln(2 + |x|)\}}{1 + |x|^3}, \quad x \in \mathbb{R},$$

$$\sup_{x \in \mathbb{R}}(1 + x^2)\Delta_n(x) \leqslant A_6 \cdot \frac{\beta_{3,1}}{\sigma_1^3 \sqrt{n}}.$$

The results of [12, 29] were further reinforced by Nagaev [33] (in the i.i.d. case with $\delta = 1$) and Bikelis [5] (in the non-i.i.d. case with arbitrary $0 < \delta \leqslant 1$), who used the methods of the theory of large deviations to prove existence of such absolute constants $K_0(\delta)$ that

$$\sup_x \left(1 + |x|^{2+\delta}\right)\Delta_n(x) \leqslant K_0(\delta)\ell_n, \quad n \in \mathbb{N}, \quad F_1, \ldots, F_n \in \mathcal{F}_{2+\delta}.$$

$$(3.1.8)$$

To be more precise, in [5] Bikelis obtained a nonuniform extension of Osipov's inequality (3.1.4) in the form

$$\Delta_n(x) \leqslant \frac{A_7}{(1 + |x|)^3 B_n^3} \int_0^{(1+|x|)B_n} L_n(z)\mathrm{d}z$$

$$= A_7 \sum_{k=1}^n \left[\frac{\mathsf{E}X_k^2 \mathbf{1}(|X_k| > (1 + |x|)B_n)}{(1 + |x|)^2 B_n^2} + \frac{\mathsf{E}|X_k|^3 \mathbf{1}(|X_k| \leqslant (1 + |x|)B_n)}{(1 + |x|)^3 B_n^3}\right]$$

$$(3.1.9)$$

for all $n \in \mathbb{N}$ and $F_1, \ldots, F_n \in \mathcal{F}_2$ with some absolute constant A_7, which trivially yields (3.1.8) with

$$K_0(\delta) = A_7 \cdot \sup_{x>0} \frac{1 + x^{2+\delta}}{(1 + x)^{2+\delta}} = A_7 \quad \text{for every } \delta \in [0, 1].$$

It is worth mentioning that in 2001 Chen and Shao [8] reproved inequality (3.1.9) (still with an unknown constant) by Stein's method.

In 1977 Ahmad and Lin [1] proved an estimate

$$\Delta_n(x) \leqslant A_8 \cdot \sum_{k=1}^n \frac{\mathsf{E}X_k^2 g(X_k)}{B_n^2 g((1 + |x|^3)B_n)}, \quad n \in \mathbb{N}, \quad F_1, \ldots, F_n \in \mathcal{F}_2, \quad g \in \mathcal{G},$$

with some absolute constant A_8, which generalizes the Katz–Petrov inequality (3.1.2), since $g((1 + |x|^3)B_n) \geqslant g(B_n)$ for $g \in \mathcal{G}$, and also yields (3.1.8) with $\delta = 1$. In 1979 Petrov [56] deduced a nonuniform analogue of his inequality (3.1.2)

$$\Delta_n(x) \leqslant A_9 \cdot \sum_{k=1}^{n} \frac{\mathsf{E}X_k^2 g(X_k)}{(1+|x|)^2 B_n^2 g((1+|x|)B_n)}, \quad n \in \mathbb{N}, \ g \in \mathcal{G},$$

(3.1.10)

with $A_9 = A_7$ by noting that the right-hand side of Bikelis' estimate (3.1.9) does not exceed the right-hand side of (3.1.10) for every $g \in \mathcal{G}$. On the other hand, inequality (3.1.9) follows from (3.1.10) with $g(u) = \min\left\{1, \frac{|u|}{(1+|x|)B_n}\right\} \in \mathcal{G}$. Moreover, inequality (3.1.10) also reinforces the above mentioned result of [1]. Upper bounds for the constant A_7 were considered in papers [24, 34, 35, 73]. The best known upper bounds for A_7 are obtained in [24]: $A_7 \leqslant 39.32$ in the i.i.d. case and $A_7 \leqslant 47.65$ in the general situation. Moreover, in [24] it is also demonstrated that for $|x| \geqslant 10$ we have $A_7 \leqslant 24.13$ in the i.i.d. case and $A_7 \leqslant 29.62$ in the general situation.

The first upper bounds for the absolute constants $K_0(\delta)$ were obtained by Paditz [45–48]. In particular, in his first work on this topic [47] which was published only in 1978, Paditz obtained an estimate for $K_0(1)$ which exceeded 1955. Later he proved [48] the bounds

$$K_0(0.9) \leqslant 820.4, \quad K_0(0.7) \leqslant 569.5, \quad K_0(0.5) \leqslant 376.7,$$
$$K_0(0.3) \leqslant 241.4, \quad K_0(0.1) \leqslant 151.3.$$

In his dissertation [46], Paditz demonstrated that $K_0(1) \leqslant 114.7$. Michel [31] showed that in the i.i.d. case $K_0(1) \leqslant C_0(1) + 8(1 + e)$. With the account of the upper bound $C_0(1) \leqslant 0.469$ [70] this inequality yields $K_0(1) \leqslant 30.22$. In his dissertation [74], Tysiak obtained the estimates

$$K_0(1.0) \leqslant 32.88, \quad K_0(0.9) \leqslant 29.83, \quad K_0(0.8) \leqslant 27.21, \quad K_0(0.7) \leqslant 25.06,$$
$$K_0(0.6) \leqslant 23.41, \quad K_0(0.5) \leqslant 21.94, \quad K_0(0.4) \leqslant 20.58, \quad K_0(0.3) \leqslant 19.32,$$
$$K_0(0.2) \leqslant 18.17, \quad K_0(0.1) \leqslant 17.05.$$

Mirakhmedov [32] stated that Michel's estimate $K_0(1) \leqslant C_0(1) + 8(1 + e)$ held true in the non-i.i.d. case as well. However, the computations in [32, 74] contained some inaccuracies (see remarks in [52, 53]). Later, Paditz and Mirakhmedov [53] announced a corrected estimate $K_0(1) \leqslant 32.153$. In 1989, Paditz [52] provided an analytical representation and described an algorithm of computation of $K_0(\delta)$ for every $\delta \in (0, 1]$ and, in particular, obtained an upper bound $K_0(1) \leqslant 31.935$.

Recently, an interest to the problem of estimation of the constants $K_0(\delta)$ rose again which lead to publication of a series of works [16, 17, 36, 38, 70], where by use of a new inequality (3.1.6) the upper bounds for the constants

$K_0(\delta)$ were improved substantially. In particular, for $\delta = 1$ there were given the estimates $K_0(1) \leqslant 18.12$ in the i.i.d. case [38] and $K_0(1) \leqslant 22.25$ in the non-i.i.d. case [16, 17].

Nonuniform analogues of the uniform estimates with an improved structure like (3.1.6) were obtained in the i.i.d. case by Gavrilenko [15] with $\delta = 1$ and by Nefedova and Shevtsova [37] with arbitrary $\delta \in (0, 1]$. In particular, in [37] there was announced an estimate

$$\sup_{x \in \mathbb{R}}(1 + |x|^3)\Delta_n(x) \leqslant 15.77 \cdot \frac{\beta_{3,1} + \sigma_1^3}{\sigma_1^3 \sqrt{n}} = 15.77(\ell_n + \tau_n),$$

$$F_1 = \cdots = F_n \in \mathcal{F}_3, \; n \in \mathbb{N},$$

which was sharper than the classical Nagaev–Bikelis bound (3.1.8) with the best known value of $K_0(1) = 18.12$ [38] for distributions with large values of the third normalized absolute moment $\beta_{3,1}/\sigma_1^3$, due to the less factor of the Lyapunov fraction ℓ_n (recall that $\ell_n \geqslant \tau_n$).

However, as it was noticed by Pinelis [59], the cited works [16, 17, 36, 38, 52, 53] contained inaccuracies, so that "the best possibly correct" upper bound for $K_0(1)$ in the non-i.i.d. case seemed to be 114.7 [46].

In paper [70] for $\delta = 1$ and in the present study for arbitrary $\delta \in (0, 1]$ the mentioned inaccuracies are corrected, and by use of inequalities (3.1.5) and (3.1.6), here we give an analytical representation and describe an algorithm of computation of the absolute constants $K_s(\delta)$ for arbitrary $s \in [0, 1]$ and $\delta \in (0, 1]$ such that

$$\sup_{x \in \mathbb{R}}(1 + |x|^{2+\delta})\Delta_n(x) \leqslant \min_{s \in [0,1]} K_s(\delta)(\ell_n + s\tau_n),$$

$$F_1, F_2, \ldots, F_n \in \mathcal{F}_{2+\delta}, \; n \in \mathbb{N}. \qquad (3.1.11)$$

The values of the constants $K_s(\delta)$ for some $s \in [0, 1]$ and $\delta \in (0, 1]$ obtained in the present study are given in Table 3.3 (for $s > s_1(\delta)$ we put $K_s(\delta) = K_{s_1}(\delta)$). The case $0 < s < 1$ is considered here for the first time. In particular, for $\delta = 1$ the following estimates are obtained: in the i.i.d. case

$$\sup_{x \in \mathbb{R}}(1 + |x|^3)\Delta_n(x) \leqslant \min\{17.36\ell_n, \; 15.70(\ell_n + 0.646\tau_n)\}$$

$$\leqslant \begin{cases} 17.36\ell_n, & \ell_n/\tau_n < 6.07, \\ 15.70(\ell_n + 0.646\tau_n), & \ell_n/\tau_n \geqslant 6.07. \end{cases}$$

Table 3.3 Upper Bounds for the Constants $K_s(\delta)$ in (3.1.11) for Some $s \in [0, 1]$ and $\delta \in (0, 1)$ Obtained in Corollary 3.4.5

	Non-i.i.d. Case			I.i.d. Case		
δ	$K_0(\delta)$	$K_{s_1}(\delta)$	s_1	$K_0(\delta)$	$K_{s_1}(\delta)$	s_1
1	21.82	18.19	1	17.36	15.70	0.646
0.9	20.07	16.65	1	16.24	14.61	0.619
0.8	18.53	15.34	1	15.20	13.61	0.625
0.7	17.14	14.20	1	14.13	12.71	0.570
0.6	15.91	13.19	0.859	13.15	11.90	0.498
0.5	14.84	12.30	0.834	12.26	11.17	0.428
0.4	13.92	11.53	0.806	11.43	10.51	0.350
0.3	13.10	10.86	0.778	10.66	9.93	0.273
0.2	12.35	10.28	0.748	9.92	9.42	0.183
0.1	11.67	9.77	0.710	9.18	8.97	0.074

and in the non-i.i.d. case

$$\sup_{x \in \mathbb{R}} (1 + |x|^3) \Delta_n(x) \leqslant \min \left\{ 21.82 \ell_n, \; 18.19(\ell_n + \tau_n) \right\}$$

$$\leqslant \begin{cases} 21.82 \ell_n, & \ell_n / \tau_n < 5.01, \\ 18.19(\ell_n + \tau_n), & \ell_n / \tau_n \geqslant 5.01, \end{cases}$$

The problem of studying the dependence of the remainder term $\Delta_n(x)$ on n and x, of course, could not be left out of the attention of Kolmogorov, who introduced [20] the functions

$$D_*(x, \delta) = \limsup_{\ell \to 0} \; \sup_{n \geqslant 1, \, F_1, \ldots, F_n \in \mathcal{F}_{2+\delta} : \, \ell_n = \ell} \Delta_n(x)/\ell,$$

$$D^*(x, \delta) = \sup_{n \geqslant 1, \, F_1, \ldots, F_n \in \mathcal{F}_{2+\delta}} \Delta_n(x)/\ell_n, \quad x \in \mathbb{R},$$

for $\delta = 1$ and posed a problem of evaluation of $D_*(x, \delta)$ and $D^*(x, \delta)$. It is easy to see that for every $0 < \delta \leqslant 1$

$$D_*(x, \delta) \leqslant D^*(x, \delta), \; x \in \mathbb{R}, \quad \sup_{x \in \mathbb{R}} |x|^{2+\delta} D^*(x, \delta) \leqslant K_0(\delta),$$

so that, inequality (3.1.8) yields $D^*(x, \delta) = O(|x|^{-2-\delta})$, $x \to \infty$. The question on the exactness of the estimates for $D^*(x, \delta)$ implied by (3.1.8) with respect to n and x was studied by Osipov and Petrov [44], Bikelis [6], Heyde [18], Michel [30], Maejima [28], Petrov [58], and Rozovsky [63]. In 1990, Chistyakov [9] managed to prove that for $\delta = 1$

$$\lim_{|x|\to\infty} |x|^3 D_*(x,1) := \lim_{|x|\to\infty} \limsup_{\ell\to 0} \sup_{n\geqslant 1,\, F_1,\ldots,F_n\in\mathcal{F}_3\,:\,\ell_n=\ell} |x|^3\Delta_n(x)/\ell = 1,$$

(3.1.12)

where the least upper bound is attained on a sequence of identical distributions. Chistyakov's result (3.1.12) also yields a lower bound $K_0(1) \geqslant 1$. In 2013, Pinelis [59] improved this lower bound to

$$K_0(1) \geqslant \sup_{F_1\in\mathcal{F}_3,\, x\in\mathbb{R}} |x|^3\Delta_1(x)\frac{\sigma_1^3}{\beta_{3,1}} \geqslant 1.0135\ldots,$$

by consideration of the Bernoulli distribution with parameter $p = 0.08$ and letting $x \to 1 - p$. However, as it will be proved below (see also [41] for the case $\delta = 1$), for every $0 < \delta \leqslant 1$

$$\sup_{n\geqslant 1,\, F_1,\ldots,F_n\in\mathcal{F}_{2+\delta}} \limsup_{|x|\to\infty} |x|^{2+\delta}\Delta_n(x)/\ell_n \leqslant 1.$$

In 1989, Nikulin in his abstract [39] (see also [40, 42]) proposed a modification of Paditz' method, which allows to replace the absolute constant $K_0(1)$ in (3.1.8) in the i.i.d. case with $\delta = 1$ by a nonincreasing function $C_N(t)$ such that for every $t \geqslant 0$

$$\sup_{|x|\geqslant t} |x|^3\Delta_n(x) \leqslant C_N(t)\cdot\ell_n, \quad F_1 = \cdots = F_n \in \mathcal{F}_3,\ n \geqslant 1,$$

and also computed the values of $C_N(t)$ for some t. In particular, he showed that $\lim_{t\to\infty} C_N(t) \leqslant 1 + e$. The latest fact allows to conclude that

$$\limsup_{|x|\to\infty} \sup_{n\geqslant 1,\, F_1=\cdots=F_n\in\mathcal{F}_3} |x|^3\Delta_n(x)/\ell_n \leqslant 1 + e < 3.72,$$

whence, in addition to (3.1.12), it follows that $\limsup_{|x|\to\infty} |x|^3 D^*(x,1) \leqslant 1 + e$ in the i.i.d. case.

In 2010 Nikulin [41] sharpened $C_N(t)$ for finite t, in particular, he showed that $C_N(3.18) \leqslant 28.41$, $C_N(5) \leqslant 16.03$, $C_N(8) \leqslant 7.26$, $C_N(10) \leqslant 5.74$. In the same paper [41] the asymptotic behavior of $C_N(x)$ as $x \to \infty$ was investigated for every *fixed* $n \geqslant 1$ and it was shown that in this situation $\lim_{x\to\infty} C_N(x) = 1$, that is,

$$\sup_{n\geqslant 1,\, F_1=\cdots=F_n\in\mathcal{F}_3} \limsup_{|x|\to\infty} |x|^3\Delta_n(x)/\ell_n \leqslant 1.$$

In papers [37, 38] it was stated that

$$\limsup_{|x|\to\infty} \sup_{n\geqslant 1,\, F_1=\cdots=F_n\in\mathcal{F}_{2+\delta}} |x|^{2+\delta}\Delta_n(x)/\ell_n \leqslant 1, \quad 0<\delta\leqslant 1,$$

that is, that in the i.i.d. case $\limsup\limits_{|x|\to\infty} |x|^{2+\delta}D^*(x,\delta) = 1$. However, the proof of this statement contains inaccuracies (see remarks in [59]), so that the correct upper bound is $1 + e$, at the present time.

The mentioned inaccuracies were corrected in [70]. Following the reasoning of [70], in this study we give an analytical representation and describe an algorithm of computation of nonincreasing positive functions $Q_s(t,\delta)$, $Q_s^*(t,\delta) \leqslant sQ_s(t,\delta)$ of the argument $t \geqslant 0$ $(Q_0^*(t,\delta) \equiv 0)$ such that for every $0 < \delta \leqslant 1$, $t \geqslant 0$, $n \geqslant 1$, and $F_1, F_2, \ldots, F_n \in \mathcal{F}_{2+\delta}$

$$\sup_{|x|\geqslant t} |x|^{2+\delta}\Delta_n(x) \leqslant \min_{0\leqslant s\leqslant 1} (Q_s(t,\delta)\ell_n + Q_s^*(t,\delta)\tau_n)$$

$$\leqslant \min_{0\leqslant s\leqslant 1} Q_s(t,\delta)(\ell_n + s\tau_n), \tag{3.1.13}$$

$$\sup_{|x|\geqslant t} |x|^{2+\delta}\Delta_n(x) \leqslant Q_0(t,\delta)\ell_n. \tag{3.1.14}$$

The constructed functions satisfy

$$\lim_{x\to\infty} Q_s(x,\delta) = 1 + e = 3.7182\ldots, \qquad \lim_{x\to\infty} Q_s^*(x,\delta) = 0,$$

$$\delta \in (0,1], \; s \in [0,1].$$

Estimates (3.1.13) and (3.1.14) in the non-i.i.d. case and with arbitrary $s \in [0,1]$ are considered in the present work for the first time.

Inequality (3.1.13) yields the following upper bounds for the Kolmogorov functions

$$\sup_{|x|\geqslant t} |x|^{2+\delta}D^*(x,\delta) \leqslant Q_0(t,\delta), \quad t \geqslant 0,$$

$$\limsup_{|x|\to\infty} |x|^{2+\delta}D^*(x,\delta) \leqslant 1 + e, \quad \delta \in (0,1].$$

From [59] it also follows that $\sup_{|x|\geqslant t} |x|^3 D^*(x,1) > 1.0135$ for $0 \leqslant t < 0.02$. The values of $Q_s(t,\delta)$ for some $t > t_s(\delta)$, $t = 0$, and $\delta \in (0,1]$ are given in Table 3.6 for $s = 0$ and in Table 3.7 for $s = s_1(\delta)$. In particular, with $\delta = 1$ and $s = 0$ the following estimates hold:

$$\sup_{|x|\geqslant t} |x|^3 \Delta_n(x) \leqslant \begin{cases} 21.26\ell_n, & t \geqslant 0, \\ 17.19\ell_n, & t \geqslant 4, \\ 12.35\ell_n, & t \geqslant 5, \\ 7.36\ell_n, & t \geqslant 10, \end{cases} \quad F_1, F_2, \ldots, F_n \in \mathcal{F}_3, \; n \in \mathbb{N},$$

$$\sup_{|x|\geqslant t} |x|^3 \Delta_n(x) \leqslant \begin{cases} 16.90\ell_n, & t \geqslant 0, \\ 14.58\ell_n, & t \geqslant 4, \\ 11.56\ell_n, & t \geqslant 5, \\ 5.85\ell_n, & t \geqslant 10, \end{cases} \quad F_1 = \cdots = F_n \in \mathcal{F}_3, \; n \in \mathbb{N}.$$

It is worth mentioning that the functions $Q_s(t, \delta)$ and $Q_s^*(t, \delta)$ constructed in the present work have the following properties for every $0 < \delta \leqslant 1$:

1. $Q_s(t, \delta)$ does not increase in $t \geqslant 0$ for every $s \in [0, 1]$;
2. $Q_s(0, \delta) = K_s(\delta) - C_s(\delta)$ for every $s \in [0, 1]$;
3. $Q_s(t, \delta) = Q_s(0, \delta)$ for $t \leqslant t_s(\delta)$ and $Q_s(t, \delta) < Q_s(0, \delta)$ for $t > t_s(\delta)$ (the values of $t_s(\delta)$ are given in Tables 3.4 and 3.5);
4. $Q_s(t, \delta) = Q_{s_1}(t, \delta)$ for every $s \in [s_1(\delta), 1]$ and $t \geqslant 0$;
5. $Q_s(t, \delta) \leqslant Q_r(t, \delta)$ for all $0 \leqslant r \leqslant s \leqslant 1$ and $t \geqslant 0$;
6. in the i.i.d. case $Q_0(t, 1) \leqslant C_N(t)$ for all $t \geqslant 0$;
7. usually, $Q_s^*(t, \delta)$ is substantially smaller than $sQ_s(t, \delta)$, already for the moderate $t > 0$.

In [59] Pinelis proposed a new approach to computation of the constants $K_0(1)$ based on the Prawitz smoothing inequality [61] similarly to the modern method of computation of the absolute constant $C_0(1)$ in the classical Berry–Esseen inequality. Pinelis' approach does not use the truncation techniques, which is crucial for Nagaev–Paditz method. Pinelis [59] also suggested that his approach would allow to improve the upper bounds for $K_0(1)$ in (3.1.8) considerably, but it still needs some more analytical results for practical realization and comparison with the known bounds. The same ideas based on traditional Berry–Esseen–Zolotarev smoothing inequalities are described in [7, § 18].

The method employed in the present work is based still on the Nagaev–Paditz ideas and is aimed for (i) correcting inaccuracies in the preceding works; (ii) perfecting the Nagaev–Paditz method to get as good bounds as possible; (iii) demonstration of effective application of new Berry–Esseen-type inequalities (3.1.6). The idea underlying the constructive method of proving nonuniform estimates, which was described by Paditz [52] and which allows to get numerical estimates of the absolute constants,

Table 3.4 Optimal Values of $t_0(\delta)$, $a_0(\delta)$, $b_0(\delta)$, $\gamma_0(\delta)$, That Deliver Minimum in (3.4.27), and Corresponding Values of $Q_0(\delta)$ From Theorem 3.4.4 and $K_0(\delta) = Q_0(\delta) + C_0(\delta)$ From Corollary 3.4.5

δ	$t_0(\delta)$	$a_0(\delta)$	$b_0(\delta)$	$\gamma_0(\delta)$	$Q_0(\delta)$	$K_0(\delta)$
Non-i.i.d. Case With $s = 0$						
1.0	3.3640	11.5566	1.6357	0.4985	21.26	21.82
0.9	3.3953	10.3229	1.6293	0.4923	19.51	20.07
0.8	3.4188	9.2513	1.6226	0.4837	17.95	18.53
0.7	3.4355	8.3069	1.6153	0.4750	16.55	17.14
0.6	3.4410	7.4937	1.6076	0.4670	15.30	15.91
0.5	3.4330	6.8019	1.5999	0.4595	14.19	14.84
0.4	3.4090	6.2205	1.5925	0.4521	13.23	13.92
0.3	3.3762	5.7090	1.5849	0.4446	12.35	13.10
0.2	3.3397	5.2403	1.5768	0.4367	11.54	12.35
0.1	3.2995	4.8110	1.5680	0.4283	10.79	11.67
I.i.d. Case With $s = 0$						
1.0	3.3023	8.3760	1.5554	0.5234	16.90	17.36
0.9	3.2956	7.6637	1.5486	0.5190	15.75	16.24
0.8	3.3012	6.9928	1.5420	0.5139	14.68	15.20
0.7	3.3420	6.2928	1.5344	0.5082	13.61	14.13
0.6	3.3879	5.6537	1.5260	0.5021	12.62	13.15
0.5	3.4266	5.0918	1.5167	0.4956	11.72	12.26
0.4	3.4650	4.5764	1.5056	0.4858	10.88	11.43
0.3	3.4973	4.1114	1.4925	0.4763	10.09	10.66
0.2	3.5360	3.6733	1.4763	0.4671	9.34	9.92
0.1	3.5933	3.2481	1.4553	0.4579	8.60	9.18

consists in the appropriate partitioning of the real line into domains of "small", "moderate", and "large" values of x. The following partitioning is traditionally used:

(i) "small" values of x: $0 \leqslant x^2 \leqslant t^2$;
(ii) "moderate" values of x: $t^2 \leqslant x^2 \leqslant c_n(x; \delta, a, b)$;
(iii) "large" values of x: $c_n(x; \delta, a, b) \vee t^2 \leqslant x^2 < \infty$,

where $t > 0$, $a > 0$, $b > 1$ are auxiliary free parameters, $c_n(x; \delta, a, b)$ is some monotonically increasing function of x (in particular, see [52, 64]). Let

$$c_n(x; \delta, a, b) = \frac{b^2}{2(b-1)} \ln \frac{|x|^{2+\delta}}{a\ell_n}.$$

δ	$t_s(\delta)$	$a_s(\delta)$	$b_s(\delta)$	$\gamma_s(\delta)$	s	A_{21}^*	$Q_s(\delta)$	$K_s(\delta)$
Table 3.5 Optimal Values of $t_s(\delta)$, $a_s(\delta)$, $b_s(\delta)$, $\gamma_s(\delta)$, That Deliver Minimum in (3.4.26) With $s = s_1(\delta)$ Which Is the Minimal Point of Minimum of $C_s(\delta)$ (Given in the Sixth Column), and Corresponding Values of $A_{21}^*(t_s(\delta), \delta, a_s(\delta), b_s(\delta), \gamma_s(\delta), 0)$ From Theorem 3.4.1 and of the Constants $Q_s(\delta)$ From Theorem 3.4.4 and $K_s(\delta) = Q_s(\delta) + C_s(\delta)$ From Corollary 3.4.5								
Non-i.i.d. Case With $s = s_1(\delta)$								
1.0	3.8814	8.4284	1.6038	0.4950	1.000	0.210	17.88	18.19
0.9	3.9202	7.4457	1.5916	0.4884	1.000	0.126	16.34	16.65
0.8	3.9460	6.6372	1.5798	0.4822	1.000	0.077	15.02	15.34
0.7	3.9620	5.9560	1.5681	0.4762	1.000	0.048	13.86	14.20
0.6	3.9707	5.3694	1.5563	0.4700	0.859	0.030	12.83	13.19
0.5	3.9700	4.8687	1.5447	0.4636	0.834	0.019	11.92	12.30
0.4	3.9585	4.4422	1.5335	0.4568	0.806	0.012	11.12	11.53
0.3	3.9348	4.0805	1.5229	0.4495	0.778	0.008	10.41	10.86
0.2	3.8980	3.7735	1.5132	0.4415	0.748	0.005	9.79	10.28
0.1	3.8437	3.5071	1.5036	0.4287	0.710	0.004	9.22	9.77
I.i.d. Case With $s = s_1(\delta)$								
1.0	3.7030	6.9292	1.5400	0.5209	0.646	0.166	15.40	15.70
0.9	3.7535	6.2497	1.5320	0.5119	0.619	0.071	14.30	14.61
0.8	3.7898	5.6539	1.5230	0.5042	0.625	0.028	13.29	13.61
0.7	3.8134	5.1303	1.5130	0.4972	0.570	0.010	12.38	12.71
0.6	3.8244	4.6717	1.5025	0.4907	0.498	0.003	11.55	11.90
0.5	3.8227	4.2710	1.4914	0.4844	0.428	0.001	10.79	11.17
0.4	3.8081	3.9213	1.4801	0.4780	0.350	0.001	10.11	10.51
0.3	3.7812	3.6147	1.4685	0.4714	0.273	0.001	9.49	9.93
0.2	3.7398	3.3493	1.4571	0.4645	0.183	0.001	8.93	9.42
0.1	3.6830	3.1200	1.4460	0.4571	0.074	0.001	8.43	8.97

3.2 AUXILIARY STATEMENTS

The proof is based on the following truncation of the r.v.'s X_j, $j = 1, 2, \ldots, n$:

$$\overline{X_j} = X_j \mathbf{1}(|X_j| \leqslant y) = \begin{cases} X_j, & \text{if } |X_j| \leqslant y, \\ 0, & \text{otherwise}, \end{cases}$$

where $y > 0$ is a truncation parameter to be chosen later. Denote

$$F_n^y(x) = \mathsf{P}\left(\sum_{j=1}^{n} \overline{X_j} < x \right), \quad x \in \mathbb{R}.$$

For another parameter $h \geqslant 0$ introduce the following notation:

$$f_j(h) = \mathsf{E}e^{h\overline{X_j}} = \mathsf{E}\exp\{hX_j\mathbf{1}(|X_j| \leqslant y)\}, \quad j = 1, 2, \ldots .$$

First, let us provide several auxiliary statements.

Lemma 3.2.1. *For arbitrary values of the parameters $h \geqslant 0$ and $y > 0$ and for every $j = 1, 2, \ldots$ we have*

$$h\sigma_j^2 - \frac{\beta_{2+\delta, j}}{y^{1+\delta}}\left(1 + hy + \frac{(hy)^2}{2}\right) \leqslant m_{1,j} := \mathsf{E}\overline{X_j}e^{h\overline{X_j}} \leqslant h\sigma_j^2 + \frac{\beta_{2+\delta, j}}{y^{1+\delta}}\, e^{hy},$$

$$\tag{3.2.1}$$

$$\sigma_j^2 - \frac{\beta_{2+\delta, j}}{y^{\delta}}\,(1 \vee hy) \leqslant m_{2,j} := \mathsf{E}\overline{X_j}^2 e^{h\overline{X_j}} \leqslant \sigma_j^2 + \frac{\beta_{2+\delta, j}}{y^{\delta}}\, e^{hy}, \quad (3.2.2)$$

$$m_{3,j} := \mathsf{E}|\overline{X_j}|^3 e^{h\overline{X_j}} \leqslant \beta_{2+\delta, j}y^{1-\delta}e^{hy}, \quad (3.2.3)$$

$$1 + \frac{h^2\sigma_j^2}{2} - \frac{\beta_{2+\delta, j}}{y^{2+\delta}}\left(hy + \frac{(hy)^2}{2} + \frac{(hy)^3}{6}\right) \leqslant f_j(h) := \mathsf{E}e^{h\overline{X_j}}$$

$$\leqslant 1 + \frac{h^2\sigma_j^2}{2} + \frac{\beta_{2+\delta, j}}{y^{2+\delta}}\, e^{hy}, \quad (3.2.4)$$

$$f_j(h) \geqslant 1 - \frac{h\beta_{2+\delta, j}}{y^{1+\delta}}. \quad (3.2.5)$$

Proof. The proof of all the estimates except the lower bounds in (3.2.2) and (3.2.4) can be found in [38]. The lower bound for $m_{2,j}$ in (3.2.2) follows from the inequality $e^x \geqslant 1 + x$, $x \in \mathbb{R}$, with further estimation of the appearing truncated moments of the r.v. X_j:

$$m_{2,j} - \sigma_j^2 = \mathsf{E}\overline{X_j}^2 e^{h\overline{X_j}} - \mathsf{E}X_j^2 \geqslant \mathsf{E}\left(\overline{X_j}^2(1 + h\overline{X_j})\right) - \mathsf{E}X_j^2$$

$$= -\mathsf{E}X_j^2\mathbf{1}(|X_j| > y) + h\mathsf{E}X_j^3\mathbf{1}(|X_j| \leqslant y)$$

$$\geqslant -\mathsf{E}|X_j|^{2+\delta}\left(y^{-\delta}\mathbf{1}(|X_j| > y) + hy^{1-\delta}\mathbf{1}(|X_j| \leqslant y)\right)$$

$$\geqslant -\beta_{2+\delta, j}(y^{-\delta} \vee hy^{1-\delta}).$$

To prove the lower bound in (3.2.4), it suffices to notice that by inequality $e^x \geqslant 1 + x + x^2/2 + x^3/6$, $x \in \mathbb{R}$, we have

$$f_j(h) \geqslant \mathsf{E}\left(1 + h\overline{X_j} + \frac{(h\overline{X_j})^2}{2} + \frac{(h\overline{X_j})^3}{6}\right) \geqslant 1 - h\,|\mathsf{E}\overline{X_j}| + \frac{h^2}{2}\mathsf{E}\overline{X_j}^2 - \frac{h^3}{6}\mathsf{E}\,|\overline{X_j}|^3,$$

and to estimate the moments

$$|E\overline{X_j}| = |EX_j\mathbf{1}(|X_j| \leqslant y)| = |EX_j\mathbf{1}(|X_j| > y)|$$
$$\leqslant E|X_j|\mathbf{1}(|X_j| > y) \leqslant \beta_{2+\delta,j}/y^{1+\delta},$$
$$E\overline{X_j}^2 - \sigma_j^2 = -EX_j^2\mathbf{1}(|X_j| > y) \geqslant -\beta_{2+\delta,j}/y^{\delta},$$
$$E|\overline{X_j}|^3 = E|X_j|^3\mathbf{1}(|X_j| \leqslant y) \leqslant \beta_{2+\delta,j}\,y^{1-\delta}.$$

$\square$

Lemma 3.2.2 (see [62]). *For every* $r \geqslant 2 + \delta$ *and* $s \geqslant 1$

$$\sum_{j=1}^{n} \left(\frac{\sigma_j}{B_n}\right)^r \leqslant \tau_n^{r/(2+\delta)}, \quad \sum_{j=1}^{n} \left(\frac{\beta_{2+\delta,j}}{B_n^{2+\delta}}\right)^s \leqslant (\ell_n)^s.$$

The following almost evident statement is well-known in the literature (see, e.g., [55, proof of theorem 7 in Ch. 5, § 3]).

Lemma 3.2.3. *For an arbitrary value of the truncation parameter* $y > 0$ *we have*

$$\sup_{x\in\mathbb{R}} |F_n^y(x) - \overline{F}_n(x)| \leqslant \sum_{j=1}^{n} P(|X_j| > y).$$

The following lemma trivially follows from the Lagrange formula (for the complete proof see, e.g., [38]).

Lemma 3.2.4.

$1°$. *Let* $q > 0$ *and* $A > 0$. *Then*

$$\sup_{v\geqslant A} |\Phi(v) - \Phi(qv)| \leqslant \frac{1}{2}\max\left\{q^2-1, \frac{1-q^2}{q^2}\right\}\cdot(v\varphi(v))\Big|_{v=\max\{1,A\min\{1,q\}\}}.$$

$2°$. *Let* $a \in \mathbb{R}$ *and* $A > 0$. *Then*

$$\sup_{v\geqslant A} |\Phi(v + a) - \Phi(v)| \leqslant |a|\,\varphi(\min\{(A + a)_+, A\}).$$

Lemma 3.2.5 (see [4]). *For every d.f.* F *with zero mean and unit variance*

$$\sup_{x\in\mathbb{R}} |F(x) - \Phi(x)| \leqslant \sup_{x>0}\left(\Phi(x) - \frac{x^2}{1 + x^2}\right) = 0.54093654\ldots =: \varkappa.$$

(3.2.6)

Lemma 3.2.6. *For arbitrary r.v.* X *with* $\mathsf{E}|X|^3 < \infty$ *and* $a := \mathsf{E}X$, $\sigma^2 := \mathsf{E}X^2$ *we have*

$$\mathsf{E}|X - a|^3 \leqslant \mathsf{E}|X|^3 + 3|a|\sigma^2 + a^2\mathsf{E}|X| - |a|^3, \qquad (3.2.7)$$

$$\mathsf{E}|X - a|^3 \leqslant M(|a|/\sigma) \cdot \mathsf{E}|X|^3, \qquad (3.2.8)$$

where

$$M(t) = \begin{cases} \left(1 - \frac{3}{2} \cdot \frac{\sqrt{b(t)} - 1}{t^{-2} - 1}\right)^{-1}, \ b(t) = \frac{9 - 6t^2 - 2t^4}{18t^2(1 - t^2)}, & 0 < t \leqslant \sqrt{3}/2, \\ 1, & t \in \{0\} \cup (\sqrt{3}/2, 1), \\ 0, & t = 1, \end{cases}$$

with equality attained at a (sequence of) two-point distribution(s). Moreover, the function $M(t)$ *monotonically increases for* $t \leqslant t_0 := \frac{1}{6}\sqrt{3(8 - 2\sqrt{7})} = 0.4750\ldots$ *and monotonically decreases for* $t_0 \leqslant t \leqslant \sqrt{3}/2$, *so that*

$$\max_{t \in [0,1]} M(t) = M(t_0) = \frac{17 + 7\sqrt{7}}{27} < 1.3156.$$

Inequality (3.2.7) is trivial (see, e.g., [23, 38]). Inequality (3.2.8) is given in [72] and improves the similar "uniform" inequalities with universal absolute constant M in [38, 60]. However, in the present work inequality (3.2.8) is used only with the "universal" constant $M = (17 + 7\sqrt{7})/27$.

3.3 CASES (I) AND (III)—"SMALL" AND "LARGE" x

In the case (i), that is, for $0 \leqslant |x| \leqslant t$, according to (3.1.6), we have

$$|x|^{2+\delta}\Delta_n(x) \leqslant C_s(\delta)(\ell_n + s\tau_n)t^{2+\delta}, \qquad (3.3.1)$$

while in the case (iii), that is, for

$$x^2 \geqslant c_n(x; \delta, a, b) := \frac{b^2}{2(b-1)} \ln \frac{|x|^{2+\delta}}{a\ell_n}$$

the following result holds. For $a > 0, b > c \geqslant 1$ denote

$$\psi_n(x; \delta, a, b, c) = \frac{b^2}{2(b-c)} \ln \frac{|x|^{2+\delta}}{a\ell_n}.$$

Then $c_n(x; \delta, a, b) = \psi_n(x; \delta, a, b, 1)$, and the functions $c_n(x; \delta, a, b)$, $\psi_n(x; \delta, a, b, 1)$ have the same signs.

Theorem 3.3.1. *Assume that* $x \in \mathbb{R}$, $a > 0$, *and* $b > c \geqslant 1$ *are such that*

$$c_n(x; \delta, a, b) > 0 \quad and \quad x^2 \geqslant \max\{(2\pi)^{-1}, \psi_n(x; \delta, a, b, c)\}.$$

Then for arbitrary $n \in \mathbb{N}$ *and* $F_1, \ldots, F_n \in \mathcal{F}_{2+\delta}$

$$|x|^{2+\delta} \Delta_n(x) \leqslant \left(b^{2+\delta} + a \left(\frac{a\ell_n}{x^{2+\delta}} \right)^{c-1} \exp\left\{ \frac{b^{2+\delta}}{a} \right\} \right) \ell_n.$$

In particular, with $c = 1$ *and* $x^2 \geqslant \max\{(2\pi)^{-1}, c_n(x; \delta, a, b)\}$, $b > 1$, $a > 0$ *we have*

$$|x|^{2+\delta} \Delta_n(x) \leqslant P(\delta, a, b) \ell_n, \quad where \quad P(\delta, a, b) = b^{2+\delta} + a \exp\left\{ \frac{b^{2+\delta}}{a} \right\}.$$

Remark 3.3.2. It can be made sure, that the function $P(\delta, a, b)$ of the argument $a > 0$ attains its minimum value for every fixed δ and b at the point $a = b^{2+\delta}$, so that

$$\inf_{a>0, b>1} P(\delta, a, b) = \inf_{b>1} b^{2+\delta}(1 + e) = 1 + e, \quad 0 < \delta \leqslant 1.$$

Proof of Theorem 3.3.1. Without loss of generality assume that $x \geqslant 0$. Since $\Delta_n(x) = \max\left\{ (1 - \overline{F}_n(xB_n)) + (\Phi(x) - 1), (\overline{F}_n(xB_n) - 1) + (1 - \Phi(x)) \right\}$, and the quantities $\overline{F}_n(xB_n) - 1$ and $\Phi(x) - 1$ are nonpositive for every $x \in \mathbb{R}$, we have

$$\Delta_n(x) \leqslant \max\left\{ 1 - \overline{F}_n(xB_n), 1 - \Phi(x) \right\}.$$

For all $x \geqslant \max\{(2\pi)^{-1/2}, \sqrt{\psi_n(x; \delta, a, b, c)}\}$ we have

$$1 - \Phi(x) \leqslant \frac{\varphi(x)}{x} = \frac{e^{-x^2/2}}{x\sqrt{2\pi}} \leqslant e^{-x^2/2} \leqslant \left(\frac{a\ell_n}{x^{2+\delta}} \right)^{b^2/(4(b-c))}.$$

Noting that $\frac{b^2}{4(b-c)} \geqslant c$ for $b \geqslant c$ and that $a\ell_n/x^{2+\delta} \leqslant 1$ due to the condition $\psi_n(x; \delta, a, b, c) > 0$, we obtain

$$1 - \Phi(x) \leqslant \left(\frac{a\ell_n}{x^{2+\delta}} \right)^c.$$

On the other hand, for $1 - \overline{F}_n(xB_n)$ we have

$$1 - \overline{F}_n(xB_n) \leqslant 1 - F_n^y(xB_n) + \left| F_n^y(xB_n) - \overline{F}_n(xB_n) \right|$$

with the truncation parameter $y = xB_n/b$, $b > 1$. By Lemma 3.2.3 and by Markov's inequality we have

$$\left|F_n^y(xB_n) - \overline{F}_n(xB_n)\right| \leqslant \sum_{j=1}^{n} \mathsf{P}(|X_j| > y) \leqslant \sum_{j=1}^{n} \frac{b^{2+\delta}\beta_{2+\delta,\,j}}{x^{2+\delta}B_n^{2+\delta}} = \frac{b^{2+\delta}\ell_n}{x^{2+\delta}}.$$

Denote

$$h := \frac{1}{y}\ln\frac{x^{2+\delta}}{a\ell_n} = \frac{2(b-c)}{bxB_n}\,\psi_n(x;\delta,a,b,c).$$

Now using again Markov's inequality and the upper bound in (3.2.4) for $f_j(h) = \mathsf{E}e^{h\overline{X}_j}$ from Lemma 3.2.1 with the given h, we obtain

$$1 - F_n^y(xB_n) = \mathsf{P}(h\overline{X}_1 + \cdots + h\overline{X}_n \geqslant hxB_n) \leqslant e^{-hxB_n}\prod_{j=1}^{n} f_j(h)$$

$$\leqslant e^{-hxB_n}\prod_{j=1}^{n}\left(1 + \frac{h^2\sigma_j^2}{2} + \frac{\beta_{2+\delta,\,j}e^{hy}}{y^{2+\delta}}\right)$$

$$\leqslant \exp\left\{-hxB_n + \frac{h^2 B_n^2}{2} + \frac{e^{hy}}{y^{2+\delta}}\sum_{j=1}^{n}\beta_{2+\delta,\,j}\right\}$$

$$= \exp\left\{-\frac{2(b-c)}{b}\,\psi_n(x;\delta,a,b,c)\right.$$

$$\left. + \frac{2(b-c)^2}{b^2 x^2}\,\psi_n^2(x;\delta,a,b,c) + \frac{b^{2+\delta}}{a}\right\}.$$

For $x^2 \geqslant \psi_n(x;\delta,a,b,c)$ we have

$$1 - F_n^y(xB_n) \leqslant \exp\left\{-\frac{2c(b-c)}{b^2}\,\psi_n(x;\delta,a,b,c) + \frac{b^{2+\delta}}{a}\right\}$$

$$= \left(\frac{a\ell_n}{x^{2+\delta}}\right)^c \exp\left\{\frac{b^{2+\delta}}{a}\right\},$$

whence it can easily be seen that the obtained majorant for $1 - F_n^y(xB_n)$ is no less than the majorant for $1 - \Phi(x)$. Summing the obtained estimates for $1 - F_n^y(xB_n)$ and $\left|F_n^y(xB_n) - \overline{F}_n(xB_n)\right|$ we arrive at the statement of the theorem. $\qquad\square$

3.4 CASE (II)—"MODERATE" x

First of all, note that for $x^2 \leqslant c_n(x; \delta, a, b)$ we have (see (ii))

$$\ell_n \leqslant \frac{x^{2+\delta}}{a} \exp\left\{-\frac{2(b-1)}{b^2} x^2\right\} =: L(x). \qquad (3.4.1)$$

Without loss of generality assume that $x > 0$. Redefine the parameters

$$y = \gamma x B_n, \quad h = \frac{(1-\gamma)x}{B_n}, \quad \gamma \in (0,1).$$

By $S_n^* = X_1^* + \cdots + X_n^*$ denote a sum of independent r.v.'s X_j^* with the d.f.'s

$$P(X_j^* < u) = \frac{1}{f_j(h)} \int_{-\infty}^{u} e^{ht}\, dP(\overline{X}_j < v), \quad u \in \mathbb{R}, \ j = 1, 2, \ldots$$

Note that

$$\mathsf{E}(X_j^*)^r = \frac{\mathsf{E}\overline{X}_j^r e^{h\overline{X}_j}}{f_j(h)} = \frac{m_{r,j}}{f_j(h)}, \quad r = 1, 2, \quad \mathsf{E}|X_j^*|^3 = \frac{m_{3,j}}{f_j(h)}.$$

It is easy to check that

$$1 - \Phi(x) = \exp\left\{\frac{h^2 B_n^2}{2}\right\} \int_x^{+\infty} e^{-hB_n v}\, d\Phi(v - hB_n), \qquad (3.4.2)$$

$$1 - F_n^y(xB_n) = \prod_{j=1}^{n} f_j(h) \int_x^{+\infty} e^{-hB_n v}\, dP(S_n^* < vB_n), \qquad (3.4.3)$$

so that with the account of Lemma 3.2.3 we have

$$\begin{aligned}
\Delta_n(x) &= \left|1 - \Phi(x) + \left(\overline{F}_n(xB_n) - F_n^y(xB_n)\right) + F_n^y(xB_n) - 1\right| \\
&\leqslant \left|\overline{F}_n(xB_n) - F_n^y(xB_n)\right| + \left|1 - \Phi(x) + F_n^y(xB_n) - 1\right| \\
&\leqslant \sum_{j=1}^{n} \mathsf{P}(|X_j| > y) + \left|\left(\exp\left\{\frac{h^2 B_n^2}{2}\right\} - \prod_{j=1}^{n} f_j(h)\right)\right. \\
&\quad \times \left. \int_x^{+\infty} e^{-hB_n v}\, dP(S_n^* < vB_n)\right| \\
&\quad + \left|\exp\left\{\frac{h^2 B_n^2}{2}\right\} \int_x^{+\infty} e^{-hB_n v}\, d\left(\Phi(v - hB_n) - P(S_n^* < vB_n)\right)\right|.
\end{aligned}$$

Applying Markov's inequality to the first term and integrating by parts the second one we obtain

$$\Delta_n(x) \leqslant \frac{\ell_n}{\gamma^{2+\delta} x^{2+\delta}} + \left| \prod_{j=1}^{n} f_j(h) - e^{h^2 B_n^2/2} \right| e^{-hxB_n} \mathsf{P}\big(S_n^* \geqslant xB_n\big)$$

$$+ 2 \exp\big\{h^2 B_n^2/2 - hxB_n\big\} \cdot \sup_{u \geqslant x} \big|\mathsf{P}(S_n^* < uB_n) - \Phi(u - hB_n)\big|$$

$$= \frac{\ell_n}{\gamma^{2+\delta} x^{2+\delta}} + I_1 \cdot I_2 + 2 \exp\big\{ -(1-\gamma^2)x^2/2\big\} I_3,$$

where

$$I_1 := \left| \prod_{j=1}^{n} f_j(h) - e^{h^2 B_n^2/2} \right| \exp\big\{ -hxB_n\big\}, \quad I_2 := \mathsf{P}\big(S_n^* \geqslant xB_n\big),$$

$$I_3 := \sup_{u \geqslant x} \big|\mathsf{P}(S_n^* < uB_n) - \Phi(u - hB_n)\big|.$$

We will estimate I_1 by a quantity $(J_1(x)\ell_n + J_1^*(x)\tau_n)x^{-2-\delta}$, I_2 by $(J_2(x) + J_2^*(x)\tau_n)$, and I_3 by $(J_3(x)\ell_n + J_3^*(x)\tau_n)x^{-2-\delta} \exp\big\{(1-\gamma^2)x^2/2\big\}$, where $J_k(x)$, $J_k^*(x)$, $k = 1, 3$, $J_2(x)$, $J_2^*(x)L(x)$ are some nonnegative functions, which do not increase for $x \geqslant t$ and may also depend on δ and on the parameters a, b, γ.

Let us formulate a statement that will be multiply used below: the function $g(x) := x^r \exp\{-sx^2\}$ is decreasing either for $x \geqslant \sqrt{r/(2s)}$, if $r > 0$, or for all $x \geqslant 0$, if $r \leqslant 0$, and hence, if $x \geqslant t \geqslant 0$, then either for $t \geqslant \sqrt{r/(2s)}$ and $r > 0$, or for $r \leqslant 0$ and all $t \geqslant 0$ we have

$$x^r \exp\{-sx^2\} \leqslant t^r \exp\{-st^2\}, \quad s > 0. \tag{3.4.4}$$

In what follows we assume that the parameters $b > 1$, $\gamma \in (0, 1)$, and $t > 0$ satisfy the following conditions:

$$2(b-1)/b^2 > \gamma(1-\gamma), \tag{3.4.5}$$

$$t^2 \geqslant b^2/(b-1), \tag{3.4.6}$$

$$t^2 \geqslant 1.5\left[2(b-1)/b^2 - \gamma(1-\gamma)\right]^{-1}, \tag{3.4.7}$$

$$t^2 \geqslant 2(1-\gamma)^{-2}, \tag{3.4.8}$$

$$t^2 \geqslant 6 + \delta/(1-\gamma^2), \tag{3.4.9}$$

and, in addition, in the non-i.i.d. case:

$$t^2 \geqslant \frac{(2+\delta)b^2}{2(2-\delta)(b-1)}, \qquad (3.4.10)$$

$$\frac{4(b-1)}{(2+\delta)b^2} > \gamma(1-\gamma), \qquad (3.4.11)$$

or, in the i.i.d. case:

$$t^2 \geqslant \frac{2(1-\delta)}{\delta} \left[\frac{4(b-1)}{\delta b^2} - \gamma(1-\gamma) \right]^{-1}. \qquad (3.4.12)$$

Observe that condition (3.4.11) in the non-i.i.d. case is stronger than condition (3.4.5). Positivity of the right-hand sides of inequalities (3.4.7) and (3.4.12) follows from condition (3.4.5). Moreover, conditions (3.4.6), (3.4.5) and (3.4.7), (3.4.8), (3.4.9), respectively, imply that

$$x^{2-\delta} L(x) \text{ decreases for } x \geqslant t, \qquad (3.4.13)$$

$$x^{1-\delta} e^{\gamma(1-\gamma)x^2} L(x) \text{ decreases for } x \geqslant t, \qquad (3.4.14)$$

$$x^2 e^{-(1-\gamma)^2 x^2/2} \text{ decreases for } x \geqslant t, \qquad (3.4.15)$$

$$x^{6+\delta} e^{-(1-\gamma^2)x^2/2} \text{ decreases for } x \geqslant t. \qquad (3.4.16)$$

3.4.1 Estimation of I_1

Let us bound

$$I_1 = \left| \prod_{j=1}^{n} f_j(h) - \exp\{h^2 B_n^2/2\} \right| e^{-hx B_n} = \left| \prod_{j=1}^{n} e^{-h^2 \sigma_j^2/2} f_j(h) - 1 \right| e^{-(1-\gamma^2)x^2/2}$$

from above. Prawitz [62] proved that for arbitrary $C_j \geqslant A_j > 0$, $B_j \in \mathbb{C}$ such that $|B_j| \leqslant C_j$, $j = 1, 2, \dots, n$, the inequality

$$\left| \prod_{j=1}^{n} B_j - \prod_{j=1}^{n} A_j \right| \leqslant \frac{1}{2} \left(\prod_{j=1}^{n} C_j + \prod_{j=1}^{n} A_j \right) \cdot \sum_{j=1}^{n} \frac{|B_j - A_j|}{A_j}.$$

holds. Let $A_j := 1$, $B_j := e^{-h^2 \sigma_j^2/2} f_j(h)$, then by (3.2.4) (see Lemma 3.2.1) we have

$$B_j \leqslant e^{-h^2 \sigma_j^2/2} \left(1 + \frac{h^2 \sigma_j^2}{2} + \frac{\beta_{2+\delta,j}}{y^{2+\delta}} e^{hy} \right) \leqslant 1 + \frac{\beta_{2+\delta,j}}{y^{2+\delta}} e^{hy - h^2 \sigma_j^2/2}$$

$$\leqslant \exp \left\{ \frac{\beta_{2+\delta,j}}{y^{2+\delta}} e^{hy} \right\} =: C_j,$$

obviously, $C_j \geqslant 1$. Now using the Prawitz inequality with A_j, B_j, and C_j specified above, with the account of $\ell_n \leqslant L(x)$ (see (3.4.1)), we obtain

$$\left| \prod_{j=1}^{n} e^{-h^2\sigma_j^2/2} f_j(h) - 1 \right| \leqslant \frac{1}{2}\left(1 + \exp\left\{\frac{B_n^{2+\delta}}{y^{2+\delta}}\,\ell_n e^{hy}\right\}\right)$$

$$\sum_{j=1}^{n} \left| e^{-h^2\sigma_j^2/2} f_j(h) - 1 \right| \leqslant A_1(x) \sum_{j=1}^{n} \left| e^{-h^2\sigma_j^2/2} f_j(h) - 1 \right|,$$

where

$$A_1(x) := \frac{1}{2}\left(1 + \exp\left\{\frac{e^{\gamma(1-\gamma)x^2} L(x)}{(\gamma x)^{2+\delta}}\right\}\right)$$

$$= \frac{1}{2}\left(1 + \exp\left\{\frac{1}{a\gamma^{2+\delta}} \exp\left\{-\left(\frac{2(b-1)}{b^2} - \gamma(1-\gamma)\right)x^2\right\}\right\}\right)$$

$$\leqslant A_1(t)$$

for $x \geqslant t$ by condition (3.4.5). Hereinafter the symbols $A(x)$, $A_j(x)$, $A_j^*(x)$, $\widehat{A}_j(x)$, $\widehat{A}_j^*(x)$, $j = 1, 2, \ldots$, stand for nonnegative functions of x, also depending on δ and on the parameters a, b, γ.

Let us now construct two-sided bounds for $(e^{-h^2\sigma_j^2/2} f_j(h) - 1)$, $j = 1, \ldots, n$. As it follows from what was said above,

$$e^{-h^2\sigma_j^2/2} f_j(h) - 1 \leqslant \frac{\beta_{2+\delta, j}}{y^{2+\delta}}\, e^{hy - h^2\sigma_j^2/2} \leqslant \frac{\beta_{2+\delta, j}}{y^{2+\delta}}\, e^{hy}.$$

On the other hand, the lower bound (3.2.4) for $f_j(h)$ and an elementary inequality $e^x \leqslant 1 + x + 0.5x^2 e^x$, $x \geqslant 0$, imply that

$$e^{-h^2\sigma_j^2/2} f_j(h) - 1 = e^{-h^2\sigma_j^2/2}\left(f_j(h) - e^{h^2\sigma_j^2/2}\right)$$

$$\geqslant e^{-h^2\sigma_j^2/2}\left(1 + \frac{h^2\sigma_j^2}{2} - \frac{\beta_{2+\delta, j}}{y^{2+\delta}}\left(hy + \frac{1}{2}(hy)^2 + \frac{1}{6}(hy)^3\right)\right.$$

$$\left. - \left(1 + \frac{h^2\sigma_j^2}{2} + \frac{h^4\sigma_j^4}{8}\, e^{h^2\sigma_j^2/2}\right)\right)$$

$$= -\frac{\beta_{2+\delta, j}}{y^{2+\delta}}\, e^{-h^2\sigma_j^2/2}\left(hy + \frac{1}{2}(hy)^2 + \frac{1}{6}(hy)^3\right) - \frac{h^4\sigma_j^4}{8}.$$

Further, observe that

$$\sigma_j^4 \leq \beta_{2+\delta,j}\, \sigma_j^{2-\delta} \leq \beta_{2+\delta,j}\, B_n^{2-\delta} = \beta_{2+\delta,j} \left(\frac{y}{\gamma x}\right)^{2-\delta},$$

and hence,

$$e^{-h^2\sigma_j^2/2} f_j(h) - 1 \geq -\frac{\beta_{2+\delta,j}}{y^{2+\delta}} \left(hy + \frac{(hy)^2}{2} + \frac{(hy)^3}{6} + \frac{(hy)^4}{8(\gamma x)^{2-\delta}} \right).$$

Comparing the obtained lower and upper bounds, we conclude that for every $j = 1, \ldots, n$

$$\left| e^{-h^2\sigma_j^2/2} f_j(h) - 1 \right| \leq \frac{\beta_{2+\delta,j}}{y^{2+\delta}} \max\left\{ e^{hy},\ hy + \frac{(hy)^2}{2} + \frac{(hy)^3}{6} + \frac{(hy)^4}{8(\gamma x)^{2-\delta}} \right\}.$$

Thus, for I_1 we finally obtain

$$I_1 \leq A_1(t) e^{-(1-\gamma^2)x^2/2} \sum_{j=1}^{n} \frac{\beta_{2+\delta,j}}{y^{2+\delta}} \max\left\{ e^{hy},\ hy + \frac{(hy)^2}{2} + \frac{(hy)^3}{6} \right.$$

$$\left. + \frac{(hy)^4}{8(\gamma x)^{2-\delta}} \right\} = A_1(t) A_2(x) \ell_n / x^{2+\delta},$$

where

$$A_2(x) := \max\left\{ \frac{e^{-(1-\gamma)^2 x^2/2}}{\gamma^{2+\delta}},\ \left(\gamma^{-1-\delta}(1-\gamma)x^2 + \gamma^{-\delta}(1-\gamma)^2 \frac{x^4}{2} \right. \right.$$

$$\left. \left. + \gamma^{1-\delta}(1-\gamma)^3 \frac{x^6}{6} + (1-\gamma)^4 \frac{x^{6+\delta}}{8} \right) e^{-(1-\gamma^2)x^2/2} \right\} \leq A_2(t)$$

by condition (3.4.16).

3.4.2 Estimation of $\sup_u |P(S_n^* - \mathrm{E}S_n^* < u\sqrt{\mathrm{D}S_n^*}) - \Phi(u)|$

When estimating I_2 and I_3 we will meet the expressions like

$$I = \sup_{u \in \mathbb{R}} \left| P\left(\frac{S_n^* - \mathrm{E}S_n^*}{\sqrt{\mathrm{D}S_n^*}} < u \right) - \Phi(u) \right| = \sup_{u \in \mathbb{R}} \left| P\left(S_n^* < u \right) - \Phi\left(\frac{u - \mathrm{E}S_n^*}{\sqrt{\mathrm{D}S_n^*}} \right) \right|.$$

The aim of the present section is to obtain an estimate of the form

$$I \leq J(x)\ell_n + J^*(x)\tau_n,$$

in the case (ii), where $J(x)$, $J^*(x)$ are some nonnegative functions, such that the functions $J(x)L(x)$, $J^*(x)L(x)$, $x^{2+\delta}e^{-(1-\gamma^2)x^2/2}J(x)$, $x^{2+\delta}e^{-(1-\gamma^2)x^2/2}J^*(x)$ do not increase for $x \geqslant t$.

Since the r.v.'s X_j^* have all power-type moments, we may use the Berry–Esseen inequality (3.1.6) with $\delta = 1$, which yields

$$I \leqslant \min_{q \geqslant 0} C_q(1) \left(\sum_{j=1}^n \frac{\mathsf{E}|X_j^* - \mathsf{E}X_j^*|^3}{(\mathsf{D}S_n^*)^{3/2}} + q \sum_{j=1}^n \frac{(\mathsf{D}X_j^*)^{3/2}}{(\mathsf{D}S_n^*)^{3/2}} \right),$$

in particular, one can take $q = 1$, $C_q(1) = 0.3057$ in the non-i.i.d. case, and $q = 0.646$, $C_q(1) = 0.3031$ in the i.i.d. case (see (3.1.5)).

For the sake of conveniences of further references, first of all, observe that by estimates (3.2.5) and (3.4.1) we have

$$f_j(h) \geqslant 1 - \frac{h\beta_{2+\delta,j}}{y^{1+\delta}} = 1 - \frac{(\gamma^{-1} - 1)\beta_{2+\delta,j}}{(\gamma x)^\delta B_n^{2+\delta}} \geqslant 1 - \frac{(\gamma^{-1} - 1)\ell_n}{(\gamma x)^\delta}$$
$$\geqslant 1 - (\gamma^{-1} - 1)(\gamma x)^{-\delta}L(x) =: A_3(x) \geqslant A_3(t) \qquad (3.4.17)$$

due to (3.4.13). In the i.i.d. case one can take

$$A_3(x) = 1 - (\gamma^{-1} - 1)(\gamma x)^{-\delta}(L(x))^{1+2/\delta} \geqslant A_3(t)$$

still due to (3.4.13). Now bound $(\mathsf{E}X_j^*)^2 = (m_{1,j}/f_j(h))^2$ from above. With the account of (3.2.1), (3.4.1), and Lemma 3.2.2 we have

$$\frac{1}{B_n^2} \sum_{j=1}^n (\mathsf{E}X_j^*)^2 \leqslant \frac{1}{B_n^2 A_3^2(t)} \sum_{j=1}^n \left(h\sigma_j^2 + \frac{\beta_{2+\delta,j}e^{hy}}{y^{1+\delta}} \right)^2$$

$$= \frac{1}{A_3^2(t)} \sum_{j=1}^n \left((1-\gamma)x \cdot \frac{\sigma_j^2}{B_n^2} + \frac{e^{\gamma(1-\gamma)x^2}}{(\gamma x)^{1+\delta}} \cdot \frac{\beta_{2+\delta,j}}{B_n^{2+\delta}} \right)^2$$

$$= \frac{1}{A_3^2(t)} \left[(1-\gamma)^2 x^2 \sum_{j=1}^n \frac{\sigma_j^4}{B_n^4} + \frac{2(1-\gamma)xe^{\gamma(1-\gamma)x^2}}{(\gamma x)^{1+\delta}} \right.$$

$$\left. \times \sum_{j=1}^n \frac{\sigma_j^2}{B_n^2} \cdot \frac{\beta_{2+\delta,j}}{B_n^{2+\delta}} + \frac{e^{2\gamma(1-\gamma)x^2}}{(\gamma x)^{2+2\delta}} \sum_{j=1}^n \left(\frac{\beta_{2+\delta,j}}{B_n^{2+\delta}} \right)^2 \right]$$

$$\leq \frac{1}{A_3^2(t)}\left[(1-\gamma)^2 x^2 \tau_n^{4/(2+\delta)} + \frac{2(1-\gamma)}{\gamma^{1+\delta}} x^{-\delta} e^{\gamma(1-\gamma)x^2} \ell_n^{2/(2+\delta)+1}\right.$$
$$\left. + \frac{e^{2\gamma(1-\gamma)x^2}}{(\gamma x)^{2+2\delta}} \ell_n^2\right] \leq A_4^*(x)\tau_n + A_4(x)\ell_n,$$

where

$$A_4^*(x) := \frac{(1-\gamma)^2}{A_3^2(t)} x^2 (L(x))^{(2-\delta)/(2+\delta)},$$

$$A_4(x) := \frac{1}{A_3^2(t)}\left[\frac{2(1-\gamma)}{\gamma^{1+\delta}x^\delta} e^{\gamma(1-\gamma)x^2} (L(x))^{2/(2+\delta)} + \frac{e^{2\gamma(1-\gamma)x^2} L(x)}{(\gamma x)^{2+2\delta}}\right].$$

Observe that the function

$$A_4^*(x)L(x) = (1-\gamma)^2 A_3^{-2}(t)x^2 (L(x))^{4/(2+\delta)}$$
$$= \frac{(1-\gamma)^2 x^6}{A_3^2(t)a^{2/(2+\delta)}} \exp\left\{-\frac{8(b-1)}{(2+\delta)b^2} x^2\right\}$$

decreases for $x \geq t$, if

$$t^2 \geq \frac{3(2+\delta)b^2}{8(b-1)}.$$

However, this condition follows from (3.4.6) and (3.4.10), since

$$\frac{3(2+\delta)}{8} \leq \max\left\{1, \frac{2+\delta}{2(2-\delta)}\right\} = \begin{cases} 1, & \delta \leq 2/3, \\ \frac{2+\delta}{2(2-\delta)}, & \delta \geq 2/3. \end{cases}$$

The function

$$A_4(x)L(x) = \frac{1}{A_3^2(t)}\left[\frac{2(1-\gamma)}{\gamma^{1+\delta}x^\delta} (L(x))^{2/(2+\delta)} \cdot e^{\gamma(1-\gamma)x^2} L(x)\right.$$
$$\left. + \frac{e^{2\gamma(1-\gamma)x^2} L^2(x)}{(\gamma x)^{2+2\delta}}\right]$$

decreases for $x \geq t$ by (3.4.13) and (3.4.14). In what follows we shall also use the fact that the functions

$$x^{2+\delta} e^{-(1-\gamma^2)x^2/2} A_4^*(x) = \frac{(1-\gamma)^2 x^4 e^{-(1-\gamma^2)x^2/2}}{A_3^2(t)a^{(2-\delta)/(2+\delta)}} x^2$$
$$\times \exp\left\{-\frac{2(2-\delta)(b-1)}{(2+\delta)b^2} x^2\right\},$$

$$x^{2+\delta}e^{-(1-\gamma^2)x^2/2}A_4(x) = \frac{1}{A_3^2(t)}\left[\frac{e^{\gamma(1-\gamma)x^2}L(x)}{\gamma^{2+2\delta}x^\delta}e^{-(1-\gamma)^2x^2/2}\right.$$
$$\left.+\frac{2(1-\gamma)}{\gamma^{1+\delta}}x^2e^{-(1-\gamma)^2x^2/2}(L(x))^{2/(2+\delta)}\right]$$

decrease for $x \geqslant t$ by (3.4.16), (3.4.10) and (3.4.15), (3.4.13), (3.4.14). In the i.i.d. case, taking into account that

$$\sum_{j=1}^{n}\frac{\sigma_j^4}{B_n^4} = \frac{1}{n} = \tau_n^{2/\delta}, \quad \sum_{j=1}^{n}\frac{\sigma_j^4}{B_n^4}\cdot\frac{\beta_{2+\delta,j}}{B_n^{2+\delta}} = \frac{\ell_n}{n} = \tau_n^{2/\delta}\ell_n,$$

$$\sum_{j=1}^{n}\left(\frac{\beta_{2+\delta,j}}{B_n^{2+\delta}}\right)^2 = \frac{\ell_n^2}{n} = \tau_n^{2/\delta}\ell_n^2,$$

one can put $A_4(x) := 0$,

$$A_4^*(x) := \frac{(L(x))^{2/\delta-1}}{A_3^2(t)}\left[(1-\gamma)x + \frac{e^{\gamma(1-\gamma)x^2}L(x)}{(\gamma x)^{1+\delta}}\right]^2,$$

and observe that

$$A_4^*(x)L(x) = \frac{(x^\delta L(x))^{2/\delta}}{A_3^2(t)}\left[1-\gamma + \frac{e^{\gamma(1-\gamma)x^2}L(x)}{\gamma^{1+\delta}x^{2+\delta}}\right]^2 \leqslant A_4^*(t)L(t)$$

by (3.4.13) and (3.4.14). Decrease of the function

$$x^{2+\delta}e^{-(1-\gamma^2)x^2/2}A_4^*(x) = \frac{(L(x))^{2/\delta-1}}{A_3^2(t)}\left[(1-\gamma)x^{3+\delta}e^{-(1-\gamma^2)x^2/2}\right.$$
$$\left.+\frac{e^{\gamma(1-\gamma)x^2}L(x)}{\gamma^{1+\delta}}xe^{-(1-\gamma^2)x^2/2}\right]^2$$

for $x \geqslant t$ follows from (3.4.13), (3.4.16), and (3.4.14).

Furthermore, by virtue of (3.2.4) we have

$$(f_j(h))^{-1} \geqslant 2 - f_j(h) \geqslant 1 - \frac{h^2\sigma_j^2}{2} - \frac{\beta_{2+\delta,j}e^{hy}}{y^{2+\delta}}.$$

Thus, taking into account Lemma 3.2.2, we obtain

$$\frac{1}{B_n^2}\sum_{j=1}^{n}\frac{\sigma_j^2}{f_j(h)} \geqslant \frac{1}{B_n^2}\sum_{j=1}^{n}\sigma_j^2\left(1 - \frac{h^2\sigma_j^2}{2} - \frac{\beta_{2+\delta,j}e^{hy}}{y^{2+\delta}}\right)$$

$$\geq \sum_{j=1}^{n} \left(\frac{\sigma_j^2}{B_n^2} - \frac{(1-\gamma)^2 x^2}{2} \cdot \frac{\sigma_j^4}{B_n^4} - \frac{e^{\gamma(1-\gamma)x^2}}{(\gamma x)^{2+\delta}} \cdot \frac{\sigma_j^2}{B_n^2} \cdot \frac{\beta_{2+\delta,j}}{B_n^{2+\delta}} \right)$$

$$\geq 1 - \frac{(1-\gamma)^2 x^2}{2} \tau_n^{4/(2+\delta)} - \frac{e^{\gamma(1-\gamma)x^2}}{(\gamma x)^{2+\delta}} \ell_n^{2/(2+\delta)+1}$$

$$\geq 1 - A_5^*(x)\tau_n - A_5(x)\ell_n,$$

where

$$A_5^*(x) := (1-\gamma)^2 x^2 (L(x))^{(2-\delta)/(2+\delta)}/2,$$
$$A_5(x) := (\gamma x)^{-2-\delta} e^{\gamma(1-\gamma)x^2} (L(x))^{2/(2+\delta)}.$$

Observe that the function

$$A_5^*(x)L(x) = \frac{(1-\gamma)^2}{2} x^2 (L(x))^{4/(2+\delta)} = \frac{(1-\gamma)^2 x^6}{2a^{4/(2+\delta)}} \exp\left\{ -\frac{8(b-1)}{(2+\delta)b^2} x^2 \right\}$$

decreases for $x \geq t$ by (3.4.6) and (3.4.10) $\left(\text{since } \frac{3(2+\delta)}{8} \leq \max\left\{1, \frac{2+\delta}{2(2-\delta)}\right\}\right)$, while the function

$$A_5(x)L(x) = (\gamma x)^{-2-\delta} (L(x))^{2/(2+\delta)} \cdot e^{\gamma(1-\gamma)x^2} L(x)$$

decreases for $x \geq t$ by (3.4.13) and (3.4.14). Moreover, the functions

$$x^{2+\delta} e^{-(1-\gamma^2)x^2/2} A_5^*(x) = (1-\gamma)^2 x^4 e^{-(1-\gamma^2)x^2/2} \cdot x^\delta (L(x))^{\frac{2-\delta}{2+\delta}}/2,$$
$$x^{2+\delta} e^{-(1-\gamma^2)x^2/2} A_5(x) = x^2 e^{-(1-\gamma)^2 x^2/2} \cdot \gamma^{-2-\delta} x^{-2} (L(x))^{2/(2+\delta)}$$

decrease by (3.4.16), (3.4.10) and (3.4.15), (3.4.13), respectively. In the i.i.d. case one can put $A_5(x) := 0$,

$$A_5^*(x) := (1-\gamma)^2 x^2 (L(x))^{2/\delta-1}/2 + (\gamma x)^{-2-\delta} e^{\gamma(1-\gamma)x^2} (L(x))^{2/\delta}.$$

As this is so, we have

$$A_5^*(x)L(x) = (1-\gamma)^2 (x^\delta L(x))^{2/\delta}/2 + (\gamma x)^{-2-\delta} e^{\gamma(1-\gamma)x^2} L(x) \cdot (L(x))^{2/\delta}$$
$$\leq A_5^*(t)L(t)$$

by (3.4.13) and (3.4.14). Decrease of the function

$$x^{2+\delta} e^{-(1-\gamma^2)x^2/2} A_5^*(x) = (1-\gamma)^2 x^4 e^{-(1-\gamma^2)x^2/2} \cdot x^\delta (L(x))^{2/\delta-1}/2$$
$$+ \gamma^{-2-\delta} e^{-(1-\gamma^2)x^2/2} (L(x))^{2/\delta}$$

for $x \geq t$ follows from (3.4.16) and (3.4.13).

Finally, using the lower bound (3.2.2) for $m_{2,j} = f_j(h)\mathsf{E}(X_j^*)^2$ together with inequalities (3.4.17) and (3.4.1), we obtain

$$
\frac{\mathsf{D}S_n^*}{B_n^2} = \frac{1}{B_n^2} \sum_{j=1}^{n} \left(\mathsf{E}(X_j^*)^2 - (\mathsf{E}X_j^*)^2 \right)
$$

$$
\geqslant \frac{1}{B_n^2} \sum_{j=1}^{n} \left(\frac{\sigma_j^2 - \beta_{2+\delta,j} y^{-\delta}(1 \vee hy)}{f_j(h)} - (\mathsf{E}X_j^*)^2 \right)
$$

$$
\geqslant 1 - A_5^*(x)\tau_n - A_5(x)\ell_n - \frac{1 \vee \gamma(1-\gamma)x^2}{A_3(t)(\gamma x)^\delta} \ell_n - A_4^*(x)\tau_n - A_4(x)\ell_n
$$

$$
=: 1 - A_6(x)\ell_n - A_6^*(x)\tau_n,
$$

where

$$
A_6(x) = A_4(x) + A_5(x) + \frac{1 \vee \gamma(1-\gamma)x^2}{A_3(t)(\gamma x)^\delta}, \quad A_6^*(x) = A_4^*(x) + A_5^*(x).
$$

Observe that for the inequality

$$
A_7(x) := 1 - (A_6(x) + A_6^*(x))L(x) \geqslant A_7(t)
$$

to hold true, it is sufficient that for $x \geqslant t$ the function

$$
\frac{1 \vee \gamma(1-\gamma)x^2}{A_3(t)(\gamma x)^\delta} L(x) = \frac{x^2 \vee \gamma(1-\gamma)x^4}{A_3(t)a\gamma^\delta} \exp\left\{ -\frac{2(b-1)}{b^2} x^2 \right\},
$$

was decreasing. But this function, according to (3.4.4), is, indeed, decreasing due to (3.4.6). For the decrease of the function $x^{2+\delta} e^{-(1-\gamma^2)x^2/2} A_6(x)$ it is sufficient that for $x \geqslant t$ the function

$$
\frac{1 \vee \gamma(1-\gamma)x^2}{A_3(t)(\gamma x)^\delta} x^{2+\delta} e^{-(1-\gamma^2)x^2/2} = \frac{x^2 \vee \gamma(1-\gamma)x^4}{A_3(t)\gamma^\delta} e^{-(1-\gamma^2)x^2/2},
$$

was decreasing. But this function is, indeed, decreasing due to (3.4.16). Thus, if $A_7(t) > 0$, then we have

$$
\mathsf{D}S_n^*/B_n^2 \geqslant 1 - A_6(x)\ell_n - A_6^*(x)\tau_n \geqslant A_7(x) \geqslant A_7(t), \qquad (3.4.18)
$$

$$
\frac{B_n^2 - \mathsf{D}S_n^*}{\mathsf{D}S_n^*} \leqslant \frac{A_6(x)\ell_n + A_6^*(x)\tau_n}{A_7(t)}, \qquad (3.4.19)
$$

with the functions $x^{2+\delta} e^{-(1-\gamma^2)x^2/2} A_6(x)$, $x^{2+\delta} e^{-(1-\gamma^2)x^2/2} A_6^*(x)$ being nonincreasing for $x \geqslant t$.

Further, by Lemma 3.2.6 we have

$$\sum_{j=1}^{n} \mathsf{E}|X_j^* - \mathsf{E}X_j^*|^3$$

$$\leqslant \min\left\{\sum_{j=1}^{n} M_j\mathsf{E}|X_j^*|^3, \sum_{j=1}^{n}\left(\mathsf{E}|X_j^*|^3 + 3\mathsf{E}(X_j^*)^2|\mathsf{E}X_j^*| + \mathsf{E}|X_j^*|(\mathsf{E}X_j^*)^2\right)\right\},$$

where $M_j := M\left(|\mathsf{E}X_j^*|/\sqrt{\mathsf{E}(X_j^*)^2}\right)$ with the function $M(\cdot)$ defined in Lemma 3.2.6. In the i.i.d. case with $\sigma_j = 1$, $j = \overline{1,n}$, by (3.4.17) and by Lemma 3.2.1 (see (3.2.1) and (3.2.2)) we obtain

$$\begin{aligned}
\frac{(\mathsf{E}X_1^*)^2}{\mathsf{E}(X_1^*)^2} &= \frac{(\mathsf{E}\overline{X_1}e^{h\overline{X_1}})^2}{\mathsf{E}e^{h\overline{X_1}}\mathsf{E}\overline{X_1}^2 e^{h\overline{X_1}}} \leqslant \frac{(h + y^{-1-\delta}e^{hy}\beta_{2+\delta,1})^2}{A_3(t)[1 - (1 \vee hy)y^{-\delta}\beta_{2+\delta,1}]} \\
&= \frac{[(1-\gamma)xn^{-1/2} + (\gamma x)^{-1-\delta}e^{\gamma(1-\gamma)x^2}\ell_n n^{-1/2}]^2}{A_3(t)[1 - (1 \vee \gamma(1-\gamma)x^2)(\gamma x)^{-\delta}\ell_n]} \\
&\leqslant \frac{[(1-\gamma)x(L(x))^{1/\delta} + (\gamma x)^{-1-\delta}e^{\gamma(1-\gamma)x^2}(L(x))^{1+1/\delta}]^2}{A_3(t)[1 - (1 \vee \gamma(1-\gamma)x^2)(\gamma x)^{-\delta}L(x)]} =: a(x),
\end{aligned}$$

with $a(x) \leqslant a(t)$ for $x \geqslant t$ in view of (3.4.13) and (3.4.14). Thus, with the account of the properties of the function $M(\cdot)$ (see Lemma 3.2.6) we have for every $j = 1, \ldots, n$

$$M_j \leqslant \begin{cases} M\left(\sqrt{a(t)} \wedge \frac{1}{6}\sqrt{3(8 - 2\sqrt{7})}\right) & \text{in the i.i.d. case,} \\ \frac{17+7\sqrt{7}}{27} < 1.3156 & \text{in the non-i.i.d. case.} \end{cases}$$

However, the majorant $M(\sqrt{a(t)} \wedge \frac{1}{6}\sqrt{3(8 - 2\sqrt{7})})\mathsf{E}|X_1^*|^3$ in the i.i.d. case turns out to give not so precise estimate for $\mathsf{E}|X_1^* - \mathsf{E}X_1^*|^3$ as the expression $\mathsf{E}|X_1^*|^3 + \cdots$ which appears in the second argument of the minimum above.

Now let us estimate the sums of moments entering into the upper bound for the sum of centered third-order absolute moments by use of Lemma 3.2.1. Start with the last term. We have

$$\sum_{j=1}^{n} \mathsf{E}|X_j^*|(\mathsf{E}X_j^*)^2 \leqslant \max_{1\leqslant j\leqslant n} \mathsf{E}|X_j^*| \sum_{j=1}^{n}(\mathsf{E}X_j^*)^2 \leqslant \max_{1\leqslant j\leqslant n} \sqrt{\mathsf{E}(X_j^*)^2} \sum_{j=1}^{n}(\mathsf{E}X_j^*)^2.$$

The sum $\sum_{j=1}^{n}(EX_j^*)^2$ has already been bounded from above by $(A_4(x)\ell_n + A_4^*(x)\tau_n)B_n^2$. By (3.2.2) we have for every $j = 1, 2, \ldots, n$

$$\mathsf{E}(X_j^*)^2 = \frac{m_{2,j}}{f_j(h)} \leqslant \frac{\sigma_j^2 + \beta_{2+\delta,j}y^{-\delta}e^{hy}}{A_3(t)} = \frac{B_n^2}{A_3(t)}\left(\frac{\sigma_j^2}{B_n^2} + \frac{e^{\gamma(1-\gamma)x^2}}{(\gamma x)^\delta} \cdot \frac{\beta_{2+\delta,j}}{B_n^{2+\delta}}\right)$$

$$\leqslant \frac{B_n^2}{A_3(t)}A_8(x),$$

where

$$A_8(x) := \begin{cases} (L(x))^{2/(2+\delta)} + (\gamma x)^{-\delta}e^{\gamma(1-\gamma)x^2}L(x) & \text{in the non-i.i.d. case,} \\ (L(x))^{2/\delta}\left(1 + (\gamma x)^{-\delta}e^{\gamma(1-\gamma)x^2}L(x)\right) & \text{in the i.i.d. case,} \end{cases}$$

moreover, $A_8(x) \leqslant A_8(t)$ by (3.4.13) and (3.4.14). Hence,

$$\sum_{j=1}^{n}\mathsf{E}|X_j^*|(EX_j^*)^2 \leqslant (A_9(x)\ell_n + A_9^*(x)\tau_n)B_n^3,$$

where $A_9(x) := A_4(x)\sqrt{A_8(t)/A_3(t)}$, $A_9^*(x) := A_4^*(x)\sqrt{A_8(t)/A_3(t)}$, and the functions $A_9(x)L(x)$, $A_9^*(x)L(x)$, $x^{2+\delta}e^{-(1-\gamma^2)x^2/2}A_9(x)$, $x^{2+\delta}e^{-(1-\gamma^2)x^2/2}A_9^*(x)$ do not increase for $x \geqslant t$.

Consider the second term

$$3\sum_{j=1}^{n}\mathsf{E}\left(X_j^*\right)^2|EX_j^*| = 3\sum_{j=1}^{n}\frac{m_{2,j}}{f_j(h)} \cdot \frac{|m_{1,j}|}{f_j(h)}$$

$$\leqslant \frac{3}{A_3^2(t)}\sum_{j=1}^{n}\left(\sigma_j^2 + \frac{\beta_{2+\delta,j}e^{hy}}{y^\delta}\right)\left(h\sigma_j^2 + \frac{\beta_{2+\delta,j}e^{hy}}{y^{1+\delta}}\right)$$

$$= \frac{3B_n^3}{A_3^2(t)}\sum_{j=1}^{n}\left(\frac{\sigma_j^2}{B_n^2} + \frac{\beta_{2+\delta,j}}{B_n^{2+\delta}} \cdot \frac{e^{\gamma(1-\gamma)x^2}}{(\gamma x)^\delta}\right)$$

$$\times \left(\left(1-\gamma\right)x\frac{\sigma_j^2}{B_n^2} + \frac{e^{\gamma(1-\gamma)x^2}}{(\gamma x)^{1+\delta}} \cdot \frac{\beta_{2+\delta,j}}{B_n^{2+\delta}}\right)$$

$$= \frac{3B_n^3}{A_3^2(t)}\left((1-\gamma)x\sum_{j=1}^{n}\frac{\sigma_j^4}{B_n^4}\right.$$

$$+ \frac{e^{\gamma(1-\gamma)x^2}}{(\gamma x)^{1+\delta}}\left(\gamma(1-\gamma)x^2 + 1\right)\sum_{j=1}^{n}\frac{\sigma_j^2}{B_n^2} \cdot \frac{\beta_{2+\delta,j}}{B_n^{2+\delta}}$$

$$\left.+ \frac{e^{2\gamma(1-\gamma)x^2}}{(\gamma x)^{1+2\delta}}\sum_{j=1}^{n}\frac{\beta_{2+\delta,j}^2}{B_n^{4+2\delta}}\right) \leqslant \frac{3B_n^3}{A_3^2(t)}\left((1-\gamma)x\tau_n^{4/(2+\delta)}\right.$$

$$+ \frac{e^{\gamma(1-\gamma)x^2}}{(\gamma x)^{1+\delta}} \left(1 + \gamma(1-\gamma)x^2\right) \tau_n^{2/(2+\delta)} \ell_n$$

$$+ \frac{e^{2\gamma(1-\gamma)x^2}}{(\gamma x)^{1+2\delta}} \ell_n^2 \Bigg)$$

$$\leqslant B_n^3 \left(A_{10}^*(x)\tau_n + A_{10}(x)\ell_n\right),$$

where

$$A_{10}^*(x) := 3A_3^{-2}(t)(1-\gamma)x(L(x))^{(2-\delta)/(2+\delta)},$$

$$A_{10}(x) := \frac{3}{A_3^2(t)} \left(\frac{e^{\gamma(1-\gamma)x^2}}{(\gamma x)^{1+\delta}} \left(1 + \gamma(1-\gamma)x^2\right) (L(x))^{2/(2+\delta)} \right.$$

$$\left. + \frac{e^{2\gamma(1-\gamma)x^2} L(x)}{(\gamma x)^{1+2\delta}} \right).$$

Moreover, the functions

$$A_{10}^*(x)L(x) = 3A_3^{-2}(t)(1-\gamma)x(L(x))^{4/(2+\delta)},$$

$$A_{10}(x)L(x) = \frac{3}{A_3^2(t)} \left(\frac{1 + \gamma(1-\gamma)x^2}{(\gamma x)^{1+\delta}} e^{\gamma(1-\gamma)x^2} \cdot (L(x))^{1+2/(2+\delta)} \right.$$

$$\left. + \frac{e^{2\gamma(1-\gamma)x^2} L^2(x)}{(\gamma x)^{1+2\delta}} \right)$$

decrease for $x \geqslant t$ by (3.4.13) and (3.4.14), and the functions

$$x^{2+\delta} e^{-(1-\gamma^2)x^2/2} A_{10}^*(x) = \frac{3(1-\gamma)}{A_3^2(t)} x^{3+\delta} e^{-(1-\gamma^2)x^2/2} (L(x))^{\frac{2-\delta}{2+\delta}},$$

$$x^{2+\delta} e^{-(1-\gamma^2)x^2/2} A_{10}(x) = \frac{3}{A_3^2(t)} \left(\frac{x^{1-\delta}}{\gamma^{1+2\delta}} e^{-(1-\gamma^2)x^2/2} L(x) \right.$$

$$\left. + \frac{\left(1 + \gamma(1-\gamma)x^2\right)}{\gamma^{1+\delta}} e^{-(1-\gamma)^2 x^2/2} \cdot x(L(x))^{2/(2+\delta)} \right)$$

decrease by (3.4.16), (3.4.13) and, respectively, by (3.4.15), (3.4.6), (3.4.10) $\left(\text{with the account of } 3(2+\delta)/8 \leqslant \max\left\{1, \frac{2+\delta}{2(2-\delta)}\right\}\right)$ (3.4.14). In the non-i.i.d. case one can take $A_{10}(x) := 0$,

$$A_{10}^*(x) := \frac{3(L(x))^{2/\delta-1}}{A_3^2(t)} \left((1-\gamma)x + \frac{e^{\gamma(1-\gamma)x^2}}{(\gamma x)^{1+\delta}} \left(1 + \gamma(1-\gamma)x^2\right) L(x) \right.$$

$$\left. + \frac{e^{2\gamma(1-\gamma)x^2}}{(\gamma x)^{1+2\delta}} (L(x))^2 \right),$$

with the functions $A_{10}^*(x)L(x)$ and

$$x^{2+\delta}e^{-(1-\gamma^2)x^2/2}A_{10}^*(x) = \frac{3(L(x))^{2/\delta-1}}{A_3^2(t)}\left((1-\gamma)x^{3+\delta}e^{-(1-\gamma^2)x^2/2}\right.$$
$$+ \frac{xL(x)}{\gamma^{1+\delta}}\left(1+\gamma(1-\gamma)x^2\right)e^{-(1-\gamma)^2x^2/2}$$
$$+ \left.\frac{x^{1-\delta}}{\gamma^{1+2\delta}}e^{-(1-\gamma)^2x^2/2}\cdot e^{\gamma(1-\gamma)x^2}(L(x))^2\right)$$

being decreasing for $x \geqslant t$ due to (3.4.13), (3.4.14), and, respectively, to (3.4.16), (3.4.13), (3.4.15), (3.4.14).

Finally, let us consider the first term

$$\sum_{j=1}^n \mathsf{E}|X_j^*|^3 = \sum_{j=1}^n \frac{m_{3,j}}{f_j(h)} \leqslant \sum_{j=1}^n \frac{y^{1-\delta}e^{hy}\beta_{2+\delta,j}}{A_3(t)} = \frac{(\gamma x)^{1-\delta}}{A_3(t)}e^{\gamma(1-\gamma)x^2}\ell_n B_n^3$$
$$=: A_{11}(x)\ell_n B_n^3,$$

with

$$A_{11}(x)L(x) = \frac{\gamma^{1-\delta}x^3}{A_3(t)a}\exp\left\{-\left(\frac{2(b-1)}{b^2}-\gamma(1-\gamma)\right)x^2\right\} \leqslant A_{11}(t)L(t)$$

by (3.4.7). Moreover, for all $x > 0$ we have

$$x^{2+\delta}e^{-(1-\gamma^2)x^2/2}A_{11}(x) = \frac{\gamma^{1-\delta}x^3e^{-(1-\gamma)^2x^2/2}}{A_3(t)}$$
$$\leqslant \left.\frac{\gamma^{1-\delta}u^3e^{-(1-\gamma)^2u^2/2}}{A_3(t)}\right|_{u=x\vee\frac{\sqrt{3}}{1-\gamma}} =: \widehat{A}_{11}(x),$$

where the function $\widehat{A}_{11}(x)$ does not increase for all $x \geqslant 0$.

Thus, for every $0 \leqslant \alpha \leqslant 1$ we obtain

$$\ell_n^* := \sum_{j=1}^n \frac{\mathsf{E}|X_j^* - \mathsf{E}X_j^*|^3}{(\mathsf{D}S_n^*)^{3/2}} \leqslant A_7^{-3/2}(t)\min\left\{M \cdot A_{11}(x)\ell_n, (A_{11}(x) + A_{10}(x)\right.$$
$$\left. + A_9(x))\ell_n + (A_{10}^*(x) + A_9^*(x))(\alpha\ell_n + (1-\alpha)\tau_n)\right\}$$
$$\leqslant A_{12}(x)\ell_n + A_{12}^*(x)\tau_n, \qquad (3.4.20)$$

where

$$A_{12}(x) := A_7^{-3/2}(t)\min\left\{M \cdot A_{11}(x), A_{11}(x) + A_{10}(x) + A_9(x) + \alpha(A_{10}^*(x)\right.$$
$$\left. + A_9^*(x))\right\},$$
$$A_{12}^*(x) := (1-\alpha)A_7^{-3/2}(t)(A_{10}^*(x) + A_9^*(x)),$$

with the functions $A_{12}(x)L(x)$, $A_{12}^*(x)L(x)$, $x^{2+\delta}e^{-(1-\gamma^2)x^2/2}A_{12}^*(x)$ being nonincreasing for $x \geqslant t$, and

$$x^{2+\delta}e^{-(1-\gamma^2)x^2/2}A_{12}(x)$$
$$\leqslant A_7^{-3/2}(t)\min\left\{M\cdot\widehat{A}_{11}(x), \widehat{A}_{11}(x)\right.$$
$$\left.+ x^{2+\delta}e^{-(1-\gamma^2)x^2/2}\left(A_{10}(x)+A_9(x)+\alpha(A_{10}^*(x)+A_9^*(x))\right)\right\}$$
$$=:\widehat{A}_{12}(x)\leqslant\widehat{A}_{12}(t).$$

Also observe that the sum $A_{12}(x) + A_{12}^*(x)$ attains its minimum value at $\alpha = 1$, while the value $\alpha = 0$ minimizes the factor at the Lyapunov fraction ℓ_n in (3.4.20).

Further, using estimates (3.2.2) and (3.4.18), and the Minkowski inequality we obtain

$$\sum_{j=1}^{n}\frac{(DX_j^*)^{3/2}}{(DS_n^*)^{3/2}} \leqslant \sum_{j=1}^{n}\frac{(E(X_j^*)^2)^{3/2}}{(DS_n^*)^{3/2}} = (DS_n^*)^{-3/2}\sum_{j=1}^{n}\left(\frac{m_{2,j}}{f_j(h)}\right)^{3/2}$$

$$\leqslant \frac{A_7^{-3/2}(t)}{A_3^{3/2}(t)B_n^3}\sum_{j=1}^{n}\left(\sigma_j^2 + \frac{e^{hy}\beta_{2+\delta,j}}{y^\delta}\right)^{3/2}$$

$$= (A_3(t)A_7(t))^{-3/2}\sum_{j=1}^{n}\left(\frac{\sigma_j^2}{B_n^2} + \frac{e^{\gamma(1-\gamma)x^2}}{(\gamma x)^\delta}\frac{\beta_{2+\delta,j}}{B_n^{2+\delta}}\right)^{3/2}$$

$$\leqslant (A_3(t)A_7(t))^{-3/2}\left[\frac{e^{\gamma(1-\gamma)x^2}}{(\gamma x)^\delta}\left(\sum_{j=1}^{n}\left(\frac{\beta_{2+\delta,j}}{B_n^{2+\delta}}\right)^{3/2}\right)^{2/3}\right.$$

$$\left.+ \left(\sum_{j=1}^{n}\frac{\sigma_j^3}{B_n^3}\right)^{2/3}\right]^{3/2}$$

$$\leqslant \left(\frac{\tau_n^{2/(2+\delta)} + (\gamma x)^{-\delta}e^{\gamma(1-\gamma)x^2}\ell_n}{A_3(t)A_7(t)}\right)^{3/2}$$

$$\leqslant A_{13}(x)\ell_n + A_{13}^*(x)\tau_n, \tag{3.4.21}$$

where $A_{13}^*(x) \equiv 0$,

$$A_{13}(x) := \left(\frac{(L(x))^{2(1-\delta)/(6+3\delta)} + (\gamma x)^{-\delta}e^{\gamma(1-\gamma)x^2}(L(x))^{1/3}}{A_3(t)A_7(t)}\right)^{3/2},$$

with the function

$$A_{13}(x)L(x) = \left(\frac{(L(x))^{2/(2+\delta)} + (\gamma x)^{-\delta}e^{\gamma(1-\gamma)x^2}L(x)}{A_3(t)A_7(t)}\right)^{3/2}$$

being monotonically decreasing for $x \geqslant t$ by (3.4.13) and (3.4.14), and the function

$$x^{2+\delta}e^{-(1-\gamma^2)x^2/2}A_{13}(x) = (A_3(t)A_7(t))^{-3/2}$$
$$\left[\left(x^{2+\delta}e^{-(1-\gamma^2)x^2/2}(L(x))^{(1-\delta)/(2+\delta)}\right)^{2/3}\right.$$
$$\left.+ \gamma^{-\delta}\left(x^{4-\delta}e^{-(1-\gamma)^2x^2} \cdot e^{\gamma(1-\gamma)x^2}L(x)\right)^{1/3}\right]^{3/2}$$

being monotonically decreasing for $x \geqslant t$ by (3.4.16), (3.4.13), (3.4.15), and (3.4.14). In the i.i.d. case we have $DX_j^*/DS_n^* = 1/n$ for every $j = 1, \ldots, n$, so that

$$\sum_{j=1}^{n} \frac{(DX_j^*)^{3/2}}{(DS_n^*)^{3/2}} = \frac{1}{\sqrt{n}} = \tau_n^{1/\delta} \leqslant (L(x))^{1/\delta-1}\tau_n,$$

and hence, one can put $A_{13}(x) \equiv 0$, $A_{13}^*(x) := (L(x))^{1/\delta-1}$. The functions $A_{13}^*(x)L(x)$, $x^{2+\delta}e^{-(1-\gamma^2)x^2/2}A_{13}^*(x)$ are, obviously, decreasing for $x \geqslant t$.

Thus, by use of inequality (3.1.5) and with the account of (3.4.18), (3.4.20), and (3.4.21), we arrive at the estimate

$$I = \sup_{u\in\mathbb{R}}\left|P\left(S_n^* - ES_n^* < u\sqrt{DS_n^*}\right) - \Phi(u)\right| \leqslant A_{14}(x)\ell_n + A_{14}^*(x)\tau_n,$$
$$(3.4.22)$$

for every $0 \leqslant \alpha \leqslant 1$ and $q \geqslant 0$, where

$$A_{14}(x) := C_q(1)(A_{12}(x) + qA_{13}(x)),$$
$$A_{14}^*(x) := C_q(1)(A_{12}^*(x) + qA_{13}^*(x)).$$

Moreover, the functions $A_{14}^*(x)L(x)$, $A_{14}(x)L(x)$, $x^{2+\delta}e^{-(1-\gamma^2)x^2/2}A_{14}^*(x)$ do not increase, and

$$x^{2+\delta}e^{-(1-\gamma^2)x^2/2}A_{14}(x) \leqslant C_q(1)\left(\widehat{A}_{12}(x) + qx^{2+\delta}e^{-(1-\gamma^2)x^2/2}A_{13}(x)\right)$$
$$=: \widehat{A}_{14}(x) \leqslant \widehat{A}_{14}(t)$$

for $x \geqslant t$.

Running ahead, we should mention that in the extremal point the expression $\rho = (A_{12}(x) + A_{12}^*(x))/(A_{13}(x) + A_{13}^*(x))$ varies within the range 3.3–4.9 in the non-i.i.d. case and 16–32 in the i.i.d. case, as the structural parameter $s \in [0, s_1(\delta)]$ of the resulting estimate (3.1.11) grows, so that $q = s_1(1)$ is an optimal (i.e., minimizing the factor of ℓ_n) choice.

To be more precise, we take $q = 1$, $C_q(1) = 0.3057$ in the non-i.i.d. case, and $q = 0.646$, $C_q(1) = 0.3031$ in the i.i.d. case (see (3.1.5)). So, in what follows we assume q to be constant.

3.4.3 Estimation of I_2

Estimate the expression $I_2 = P\left(S_n^* \geqslant xB_n\right)$. With the account of (3.4.22) we have

$$I_2 \leqslant \left| P(S_n^* < xB_n) - \Phi\left(\frac{xB_n - \mathsf{E}S_n^*}{\sqrt{\mathsf{D}S_n^*}}\right) \right| + \Phi\left(-\frac{xB_n - \mathsf{E}S_n^*}{\sqrt{\mathsf{D}S_n^*}}\right)$$

$$\leqslant A_{14}(t)L(t) + A_{14}^*(x)\tau_n + \Phi\left(\frac{\mathsf{E}S_n^* - xB_n}{\sqrt{\mathsf{D}S_n^*}}\right).$$

Construct a negative upper bound for

$$\frac{\mathsf{E}S_n^* - xB_n}{\sqrt{\mathsf{D}S_n^*}} = \frac{\mathsf{E}S_n^* - hB_n^2 - \gamma xB_n}{\sqrt{\mathsf{D}S_n^*}}.$$

By use of the upper bound (3.2.1) for $m_{1,j} = \mathsf{E}X_j e^{hX_j} = f_j(h)\mathsf{E}X_j^*$, the lower bounds (3.2.5), (3.4.17) for $f_j(h)$ as well as Lemma 3.2.2, we obtain

$$\mathsf{E}S_n^* - hB_n^2 = \sum_{j=1}^n \frac{m_{1,j} - h\sigma_j^2 f_j(h)}{f_j(h)}$$

$$\leqslant \sum_{j=1}^n \frac{1}{f_j(h)}\left(h\sigma_j^2 + \frac{\beta_{2+\delta,j}e^{hy}}{y^{1+\delta}} - h\sigma_j^2 f_j(h)\right)$$

$$\leqslant \frac{1}{A_3(t)}\sum_{j=1}^n \left(h\sigma_j^2(1 - f_j(h)) + \frac{\beta_{2+\delta,j}e^{hy}}{y^{1+\delta}}\right)$$

$$\leqslant \frac{1}{A_3(t)}\sum_{j=1}^n \left(\frac{h^2\sigma_j^2\beta_{2+\delta,j}}{y^{1+\delta}} + \frac{e^{hy}\beta_{2+\delta,j}}{y^{1+\delta}}\right)$$

$$= \frac{B_n}{A_3(t)}\sum_{j=1}^n \left(\frac{(1-\gamma)^2 x^{1-\delta}}{\gamma^{1+\delta}} \cdot \frac{\sigma_j^2}{B_n^2} \cdot \frac{\beta_{2+\delta,j}}{B_n^{2+\delta}}\right.$$

$$\left. + \frac{e^{\gamma(1-\gamma)x^2}}{(\gamma x)^{1+\delta}} \cdot \frac{\beta_{2+\delta,j}}{B_n^{2+\delta}}\right)$$

$$\leqslant (A_{15}(x)\ell_n + A_{15}^*(x)\tau_n)B_n, \tag{3.4.23}$$

where

$$A_{15}(x) = \begin{cases} \dfrac{(1-\gamma)^2 x^{1-\delta}(L(x))^{\frac{2}{2+\delta}} + x^{-1-\delta} e^{\gamma(1-\gamma)x^2}}{A_3(t)\gamma^{1+\delta}} & \text{in the non-i.i.d. case,} \\ A_3^{-1}(t)(\gamma x)^{-1-\delta} e^{\gamma(1-\gamma)x^2} & \text{in the i.i.d. case,} \end{cases}$$

$$A_{15}^*(x) = \begin{cases} 0 & \text{in the non-i.i.d. case,} \\ A_3^{-1}(t)(1-\gamma)^2\gamma^{-1-\delta}x^{1-\delta}(L(x))^{2/\delta} & \text{in the i.i.d. case,} \end{cases}$$

and the functions $A_{15}^*(x)$; $A_{15}(x)L(x)$; $x^{2+\delta}e^{-(1-\gamma^2)x^2/2}A_{15}(x)$; $x^{2+\delta}e^{-(1-\gamma^2)x^2/2}A_{15}^*(x)$ do not increase for $x \geqslant t$, respectively, due to (3.4.13); (3.4.13) and (3.4.14); (3.4.16), (3.4.13); also (3.4.16), (3.4.13). Thus, for the nominator of the quantity under consideration we have

$$\mathsf{E}S_n^* - xB_n = \mathsf{E}S_n^* - hB_n^2 - \gamma x B_n \leqslant -A_{16}(x)B_n,$$

where

$$A_{16}(x) = \gamma x - (A_{15}(x) + A_{15}^*(x))L(x) \geqslant A_{16}(t).$$

In what follows assume that $A_{16}(t) > 0$.

In addition to the lower bound $\mathsf{D}S_n^* \geqslant A_7(t)B_n^2$ (see (3.4.18)) we will also need an upper bound for $\mathsf{D}S_n^*$. By use of the upper bound (3.2.2) for $m_{2,j}$, the lower bounds (3.2.5) and (3.4.17) for $f_j(h)$, and the identity $(1-z)^{-1} = (1-z)^{-1}z + 1$ we obtain

$$\mathsf{D}S_n^* \leqslant \sum_{j=1}^{n} \mathsf{E}(X_j^*)^2 = \sum_{j=1}^{n} \frac{m_{2,j}}{f_j(h)} \leqslant \sum_{j=1}^{n} \frac{\sigma_j^2 + e^{hy}y^{-\delta}\beta_{2+\delta,j}}{1 - hy^{-1-\delta}\beta_{2+\delta,j}}$$

$$= \sum_{j=1}^{n} \left(\left(1 - \frac{h\beta_{2+\delta,j}}{y^{1+\delta}}\right)^{-1}\frac{h\beta_{2+\delta,j}}{y^{1+\delta}} + 1\right)\left(\sigma_j^2 + \frac{e^{hy}\beta_{2+\delta,j}}{y^{\delta}}\right)$$

$$\leqslant \sum_{j=1}^{n} \left(\frac{h\beta_{2+\delta,j}}{A_3(t)y^{1+\delta}} + 1\right)\left(\sigma_j^2 + \frac{e^{hy}\beta_{2+\delta,j}}{y^{\delta}}\right)$$

$$= B_n^2 \sum_{j=1}^{n} \left(\frac{\sigma_j^2}{B_n^2} + \frac{e^{\gamma(1-\gamma)x^2}}{(\gamma x)^{\delta}}\cdot\frac{\beta_{2+\delta,j}}{B_n^{2+\delta}}\right.$$

$$\left. + \frac{(1-\gamma)x^{-\delta}}{A_3(t)\gamma^{1+\delta}}\cdot\frac{\sigma_j^2}{B_n^2}\cdot\frac{\beta_{2+\delta,j}}{B_n^{2+\delta}} + \frac{(1-\gamma)e^{\gamma(1-\gamma)x^2}}{A_3(t)\gamma^{1+2\delta}x^{2\delta}}\cdot\frac{\beta_{2+\delta,j}^2}{B_n^{4+2\delta}}\right)$$

$$\leqslant A_{17}(x)B_n^2 \leqslant A_{17}(t)B_n^2,$$

where

$$A_{17}(x) = 1 + \frac{e^{\gamma(1-\gamma)x^2}L(x)}{(\gamma x)^\delta} + \frac{(1-\gamma)x^{-\delta}}{A_3(t)\gamma^{1+\delta}}(L(x))^{1+2/(2+\delta)}$$
$$+ \frac{(1-\gamma)e^{\gamma(1-\gamma)x^2}}{A_3(t)\gamma^{1+2\delta}x^{2\delta}}(L(x))^2,$$

and in the i.i.d. case one can take

$$A_{17}(x) = 1 + \frac{e^{\gamma(1-\gamma)x^2}L(x)}{(\gamma x)^\delta} + \frac{(1-\gamma)x^{-\delta}}{A_3(t)\gamma^{1+\delta}}(L(x))^{1+2/\delta}$$
$$+ \frac{(1-\gamma)e^{\gamma(1-\gamma)x^2}}{A_3(t)\gamma^{1+2\delta}x^{2\delta}}(L(x))^{2+2/\delta},$$

with the inequality $A_{17}(x) \leqslant A_{17}(t)$ being true due to (3.4.14) and (3.4.13) in both cases.

Thus, under assumption that $A_{16}(t) > 0$, we have

$$\frac{xB_n - ES_n^*}{\sqrt{DS_n^*}} \geqslant \frac{A_{16}(t)}{\sqrt{A_{17}(t)}}, \tag{3.4.24}$$

and hence, $\Phi\big((ES_n^* - xB_n)/\sqrt{DS_n^*}\big) \leqslant \Phi\big(-A_{16}(t)/\sqrt{A_{17}(t)}\big)$ in view of monotonicity of the function $\Phi(u)$, $u \in \mathbb{R}$. So, under assumption that $A_{16}(t) > 0$, we finally obtain

$$I_2 \leqslant A_{14}(t)L(t) + A_{14}^*(x)\tau_n + \Phi\big(-A_{16}(t)/\sqrt{A_{17}(t)}\big).$$

3.4.4 Estimation of I_3
Let us estimate

$$I_3 = \sup_{u \geqslant x}\big|P\big(S_n^* < uB_n\big) - \Phi\big(u - hB_n\big)\big|$$

from above. We have

$$I_3 = \sup_{u \geqslant x}\bigg|P\bigg(\frac{S_n^* - ES_n^*}{\sqrt{DS_n^*}} < \frac{uB_n - ES_n^*}{\sqrt{DS_n^*}}\bigg) \pm \Phi\bigg(\frac{uB_n - ES_n^*}{\sqrt{DS_n^*}}\bigg)$$
$$\pm \Phi\bigg(\frac{uB_n - ES_n^*}{B_n}\bigg) - \Phi\big(u - hB_n\big)\bigg|$$
$$\leqslant \sup_{v \in \mathbb{R}}\big|P\big(S_n^* - ES_n^* < v\sqrt{DS_n^*}\big) - \Phi(v)\big|$$
$$+ \sup_{v \geqslant (xB_n - ES_n^*)/\sqrt{DS_n^*}}\bigg|\Phi(v) - \Phi\bigg(\frac{v\sqrt{DS_n^*}}{B_n}\bigg)\bigg|$$
$$+ \sup_{u \geqslant x}\big|\Phi\big(u - ES_n^*/B_n\big) - \Phi\big(u - hB_n\big)\big| =: I_{31} + I_{32} + I_{33}.$$

By inequality (3.4.22) we have

$$I_{31} = \sup_{v \in \mathbb{R}} \left| P\left(S_n^* - \mathrm{E}S_n^* < v\sqrt{\mathrm{D}S_n^*}\right) - \Phi(v) \right| \leqslant A_{14}(x)\ell_n + A_{14}^*(x)\tau_n.$$

For I_{32} by (3.4.24) we have

$$I_{32} = \sup_{v \geqslant (xB_n - \mathrm{E}S_n^*)/\sqrt{\mathrm{D}S_n^*}} \left| \Phi(v) - \Phi\left(\frac{v\sqrt{\mathrm{D}S_n^*}}{B_n}\right) \right|$$

$$\leqslant \sup_{v \geqslant A_{16}(t)/\sqrt{A_{17}(t)}} \left| \Phi(v) - \Phi\left(\frac{v\sqrt{\mathrm{D}S_n^*}}{B_n}\right) \right|.$$

By use of statement $1°$ of Lemma 3.2.4 under the assumption that $A_{16}(t) > 0$, as well as of the estimate $1 \wedge \sqrt{\mathrm{D}S_n^*}/B_n \geqslant 1 \wedge \sqrt{A_7(t)} = \sqrt{A_7(t)}$ (see (3.4.18)) and observing that the function $u\varphi(u)$ decreases for $u \geqslant 1$, we obtain

$$I_{32} \leqslant \max\left\{ \frac{\mathrm{D}S_n^*}{B_n^2} - 1, \frac{B_n^2 - \mathrm{D}S_n^*}{\mathrm{D}S_n^*} \right\} A_{18}(t),$$

$$A_{18}(t) = \left. \frac{u\varphi(u)}{2} \right|_{u = 1 \vee A_{16}(t)\sqrt{A_7(t)/A_{17}(t)}}.$$

In view of Lemma 3.2.1 we have

$$\frac{\mathrm{D}S_n^*}{B_n^2} - 1 = \sum_{j=1}^{n} \frac{\mathrm{D}X_j^* - \sigma_j^2}{B_n^2} \leqslant \sum_{j=1}^{n} \frac{\mathrm{E}(X_j^*)^2 - \sigma_j^2}{B_n^2}$$

$$= \sum_{j=1}^{n} \frac{m_{2,j} - \sigma_j^2 + \sigma_j^2(1 - f_j(h))}{B_n^2 f_j(h)}$$

$$\leqslant \frac{1}{B_n^2 A_3(t)} \sum_{j=1}^{n} \left(\frac{e^{hy}\beta_{2+\delta,j}}{y^{\delta}} + \frac{h\sigma_j^2 \beta_{2+\delta,j}}{y^{1+\delta}} \right)$$

$$= \frac{(\gamma x)^{-\delta}}{A_3(t)} \sum_{j=1}^{n} \left(\frac{\beta_{2+\delta,j}}{B_n^{2+\delta}} e^{\gamma(1-\gamma)x^2} + \frac{1-\gamma}{\gamma} \cdot \frac{\sigma_j^2}{B_n^2} \cdot \frac{\beta_{2+\delta,j}}{B_n^{2+\delta}} \right)$$

$$\leqslant \frac{(\gamma x)^{-\delta}\ell_n}{A_3(t)} \left(e^{\gamma(1-\gamma)x^2} + \frac{1-\gamma}{\gamma}\tau_n^{2/(2+\delta)} \right)$$

$$\leqslant A_{19}(x)\ell_n + A_{19}^*(x)\tau_n,$$

where

$$
A_{19}(x) = \begin{cases} \dfrac{e^{\gamma(1-\gamma)x^2}+(\gamma^{-1}-1)(L(x))^{2/(2+\delta)}}{A_3(t)(\gamma x)^\delta} & \text{in the non-i.i.d. case,} \\ A_3^{-1}(t)(\gamma x)^{-\delta}e^{\gamma(1-\gamma)x^2} & \text{in the i.i.d. case,} \end{cases}
$$

$$
A_{19}^*(x) = \begin{cases} 0 & \text{in the non-i.i.d. case,} \\ A_3^{-1}(t)(\gamma x)^{-\delta}(\gamma^{-1}-1)(L(x))^{2/\delta} & \text{in the i.i.d. case.} \end{cases}
$$

Notice that in both cases the functions $x^{2+\delta}e^{-(1-\gamma^2)x^2/2}A_{19}(x)$, $x^{2+\delta}e^{-(1-\gamma^2)x^2/2}A_{19}^*(x)$ do not increase for $x \geqslant t$ in view of (3.4.15), (3.4.16), and (3.4.13).

Thus, with the account of the estimate $(B_n^2 - DS_n^*)/DS_n^* \leqslant (A_6(x)\ell_n + A_6^*(x)\tau_n)/A_7(t)$ (see (3.4.19)), we conclude that

$$
I_{32} \leqslant \max\left\{ A_{19}(x)\ell_n + A_{19}^*(x)\tau_n, \frac{A_6(x)\ell + A_6^*(x)\tau_n}{A_7(t)} \right\}
$$

$$
A_{18}(t) \leqslant A_{20}(x)\ell_n + A_{20}^*(x)\tau_n,
$$

where

$$
A_{20}(x) = A_{18}(t)\cdot\max\left\{A_{19}(x)+\alpha A_{19}^*(x), (A_6(x)+\alpha A_6^*(x))/A_7(t)\right\},
$$
$$
A_{20}^*(x) = (1-\alpha)A_{18}(t)\cdot\max\left\{A_{19}^*(x), A_6^*(x)/A_7(t)\right\}, \quad 0 \leqslant \alpha \leqslant 1,
$$

and the functions $x^{2+\delta}e^{-(1-\gamma^2)x^2/2}A_{20}(x)$, $x^{2+\delta}e^{-(1-\gamma^2)x^2/2}A_{20}^*(x)$ do not increase for $x \geqslant t$. Notice that the value $\alpha = 1$ minimizes the sum $A_{20}(x)+A_{20}^*(x)$ of the factors of ℓ_n and τ_n, and the value $\alpha = 0$ minimizes the factor $A_{20}(x)$ of ℓ_n.

Finally, consider

$$
I_{33} = \sup_{u \geqslant x}\left| \Phi\left(u - \frac{ES_n^*}{B_n}\right) - \Phi(u - hB_n) \right|
$$
$$
= \sup_{v \geqslant x - hB_n}\left| \Phi\left(v + \frac{hB_n^2 - ES_n^*}{B_n}\right) - \Phi(v) \right|.
$$

Taking into account that $x - hB_n = x\gamma \geqslant \gamma t > 0$, and using statement 2° of Lemma 3.2.4, we obtain

$$
I_{33} \leqslant \frac{|hB_n^2 - ES_n^*|}{B_n} \varphi\left(\min\left\{ \left(\gamma t + \frac{hB_n^2 - ES_n^*}{B_n}\right)_+, \gamma t\right\}\right).
$$

The lower bound for $hB_n^2 - ES_n^*$ was constructed while estimation of I_2. Let us now bound this quantity from above. By use of the lower bound (3.2.1) for $m_{1,j} = E\overline{X}_j e^{h\overline{X}_j} = f_j(h)EX_j^*$ we have

$$hB_n^2 - \mathsf{E}S_n^* = \sum_{j=1}^{n} \left(h\sigma_j^2 - \frac{m_{1,j}}{f_j(h)} \right) = \sum_{j=1}^{n} \frac{h\sigma_j^2 f_j(h) \pm h\sigma_j^2 - m_{1,j}}{f_j(h)}$$

$$\leqslant \frac{1}{A_3(t)} \sum_{j=1}^{n} \left(h\sigma_j^2 (f_j(h) - 1) + \frac{\beta_{2+\delta,j}}{y^{1+\delta}} \left(1 + hy + \frac{(hy)^2}{2} \right) \right).$$

Estimate the expression $h\sigma_j^2(f_j(h) - 1)$ by inequality (3.2.4) to obtain that

$$\sum_{j=1}^{n} h\sigma_j^2 (f_j(h) - 1) \leqslant \sum_{j=1}^{n} h\sigma_j^2 \left(\frac{h^2\sigma_j^2}{2} + \frac{\beta_{2+\delta,j} e^{hy}}{y^{2+\delta}} \right)$$

$$= \frac{(hy)^3}{y^{1+\delta}} \sum_{j=1}^{n} \left(\frac{\sigma_j^4}{2y^{2-\delta}} + \frac{\sigma_j^2 \beta_{2+\delta,j} e^{hy}}{h^2 y^4} \right)$$

$$\leqslant \frac{(hy)^3}{y^{1+\delta}} \sum_{j=1}^{n} \beta_{2+\delta,j} \left(\frac{1}{2(\gamma x)^{2-\delta}} \cdot \frac{\sigma_j^{2-\delta}}{B_n^{2-\delta}} \right.$$

$$\left. + \frac{e^{\gamma(1-\gamma)x^2}}{(1-\gamma)^2 x^2 (\gamma x)^4} \cdot \frac{\sigma_j^2}{B_n^2} \right)$$

$$\leqslant (hy)^3 A(x) \sum_{j=1}^{n} \frac{\beta_{2+\delta,j}}{y^{1+\delta}},$$

where

$$A(x) = \frac{1}{2}(\gamma x)^{\delta-2}(L(x))^{(2-\delta)/(2+\delta)} + \frac{e^{\gamma(1-\gamma)x^2}}{\gamma^4(1-\gamma)^2 x^6} (L(x))^{2/(2+\delta)}$$

$$= \frac{a^{-(2-\delta)/(2+\delta)}}{2\gamma^{2-\delta}} \exp\left\{ -\frac{2(2-\delta)(b-1)}{(2+\delta)b^2} x^2 \right\}$$

$$+ \frac{a^{-2/(2+\delta)}}{\gamma^4(1-\gamma)^2 x^4} \exp\left\{ -\left(\frac{4(b-1)}{(2+\delta)b^2} - \gamma(1-\gamma) \right) x^2 \right\}.$$

Moreover, $A(x) \leqslant A(t)$ for $x^2 \geqslant t^2$ under assumption (3.4.11). In the i.i.d. case one can take

$$A(x) = \frac{1}{2}(\gamma x)^{\delta-2}(L(x))^{2/\delta-1} + \frac{e^{\gamma(1-\gamma)x^2}}{\gamma^4(1-\gamma)^2 x^6} (L(x))^{2/\delta}$$

$$= \frac{x^{4/\delta-2}}{2\gamma^{2-\delta}a^{2/\delta+1}} \exp\left\{ -\frac{2(2-\delta)(b-1)}{\delta b^2} x^2 \right\}$$

$$+ \frac{x^{4/\delta-4}a^{-2/\delta}}{\gamma^4(1-\gamma)^2} \exp\left\{ -\left(\frac{4(b-1)}{\delta b^2} - \gamma(1-\gamma) \right) x^2 \right\} \leqslant A(t)$$

due to (3.4.6) and (3.4.12). Further, under the assumption that $A(t) \leqslant 1/6$, we have

$$hB_n^2 - ES_n^* \leqslant \frac{1}{A_3(t)} \sum_{j=1}^{n} \left(\frac{(hy)^3}{6} \cdot \frac{\beta_{2+\delta,j}}{y^{1+\delta}} + \frac{\beta_{2+\delta,j}}{y^{1+\delta}} \left(1 + hy + \frac{(hy)^2}{2} \right) \right)$$

$$= \left(1 + hy + \frac{(hy)^2}{2} + \frac{(hy)^3}{6} \right) \sum_{j=1}^{n} \frac{\beta_{2+\delta,j}}{A_3(t)y^{1+\delta}} \leqslant \sum_{j=1}^{n} \frac{\beta_{2+\delta,j} e^{hy}}{A_3(t)y^{1+\delta}}.$$

On the other hand, while estimation of I_2 (see (3.4.23)) it was proved that

$$ES_n^* - hB_n^2 \leqslant \frac{1}{A_3(t)} \sum_{j=1}^{n} \left(\frac{h^2 \sigma_j^2 \beta_{2+\delta,j}}{y^{1+\delta}} + \frac{e^{hy} \beta_{2+\delta,j}}{y^{1+\delta}} \right)$$

$$\leqslant (A_{15}(x)\ell_n + A_{15}^*(x)\tau_n) B_n,$$

where the functions $A_{15}^*(x)$, $A_{15}(x)L(x)$, $x^{2+\delta} e^{-(1-\gamma^2)x^2/2} A_{15}^*(x)$, $x^{2+\delta} e^{-(1-\gamma^2)x^2/2} A_{15}(x)$ decrease for $x \geqslant t$. Comparing the previous estimates for $ES_n^* - hB_n^2$ with the just constructed upper bound for $hB_n^2 - ES_n^*$, we conclude that

$$\frac{|hB_n^2 - ES_n^*|}{B_n} \leqslant \frac{1}{A_3(t)B_n} \sum_{j=1}^{n} \left(\frac{h^2 \sigma_j^2 \beta_{2+\delta,j}}{y^{1+\delta}} + \frac{e^{hy} \beta_{2+\delta,j}}{y^{1+\delta}} \right)$$

$$\leqslant A_{15}(x)\ell_n + A_{15}^*(x)\tau_n,$$

$$\min \left\{ \left(\gamma t + \frac{hB_n^2 - ES_n^*}{B_n} \right)_+, \gamma t \right\}$$

$$\geqslant \gamma t - (A_{15}(t) + A_{15}^*(t))L(t) = A_{16}(t),$$

in view of $(A_{15}(t) + A_{15}^*(t))L(t) > 0$ and $A_{16}(t) > 0$. Since the normal density $\varphi(u)$ decreases for $u > 0$, we have

$$I_{33} \leqslant (A_{15}(x)\ell_n + A_{15}^*(x)\tau_n) \varphi(A_{16}(t)).$$

Finally, for every x from the domain (ii) under consideration we obtain that

$$|x|^{2+\delta} \Delta_n(x) \leqslant \frac{\ell_n}{\gamma^{2+\delta}} + |x|^{2+\delta} I_1 \cdot I_2 + 2|x|^{2+\delta} e^{-(1-\gamma^2)x^2/2} \cdot I_3$$

$$\leqslant \frac{\ell_n}{\gamma^{2+\delta}} + A_1(t)A_2(t)\ell_n \left(A_{14}(t)L(t) + A_{14}^*(x)\tau_n \right.$$

$$+ \Phi\left(-A_{16}(t)/\sqrt{A_{17}(t)} \right) \right)$$

$$+ 2\widehat{A}_{14}(t)\ell_n + 2t^{2+\delta}e^{-(1-\gamma^2)t^2/2}\big(A_{14}^*(t)\tau_n + A_{20}(t)\ell_n$$
$$+ A_{20}^*(t)\tau_n + (A_{15}(t)\ell_n + A_{15}^*(t)\tau_n)\varphi(A_{16}(t)))$$
$$\leqslant A_{21}(t)\ell_n + A_{21}^*(t)\tau_n, \tag{3.4.25}$$

where

$$A_{21}(x) = A_{21}(x, \delta, a, b, \gamma, \alpha) = \gamma^{-2-\delta} + 2\widehat{A}_{14}(x) + A_1(x)A_2(x)\big(A_{14}(x)L(x)$$
$$+ \Phi\big(- A_{16}(x)/\sqrt{A_{17}(x)}\,\big)\big) + 2x^{2+\delta}e^{-(1-\gamma^2)x^2/2}\,(A_{20}(x)$$
$$+ A_{15}(x)\varphi(A_{16}(x)))\,,$$
$$A_{21}^*(x) = A_{21}^*(x, \delta, a, b, \gamma, \alpha) = A_1(x)A_2(x)A_{14}^*(x)L(x)$$
$$+ 2x^{2+\delta}e^{-(1-\gamma^2)x^2/2}\,(A_{14}^*(x) + A_{20}^*(x) + A_{15}^*(x)\varphi(A_{16}(x)))\,,$$

for every $0 \leqslant \alpha \leqslant 1$, with $\alpha = 1$ minimizing the sum of the factors $A_{21}(x) + A_{21}^*(x)$ of ℓ_n and τ_n in (3.4.25), and $\alpha = 0$ minimizing the factor $A_{21}(x)$ of ℓ_n.

Summarizing the above said we arrive at the following theorem.

Theorem 3.4.1. *Assume that $t > 0$, $t^2 \leqslant x^2 \leqslant c_n(x; \delta, a, b)$, $0 < \gamma < 1$, $a > 0$, $b > 1$ are such that $A(t) \leqslant 1/6$, $A_3(t) > 0$, $A_7(t) > 0$, $A_{16}(t) > 0$, and conditions (3.4.5)–(3.4.11) in the non-i.i.d. case, or (3.4.5)–(3.4.9), (3.4.12)*** in the i.i.d. case, are fulfilled. Then for every $n \in \mathbb{N}$ and $F_1, F_2, \ldots, F_n \in \mathcal{F}_{2+\delta}$ we have*

$$|x|^{2+\delta}\Delta_n(x) \leqslant \min_{0\leqslant\alpha\leqslant1} \big(A_{21}(t, \delta, a, b, \gamma, \alpha)\ell_n + A_{21}^*(t, \delta, a, b, \gamma, \alpha)\tau_n\big).$$

Remark 3.4.2. The functions $A_{21}(t, \delta, a, b, \gamma, \alpha)\ell_n$, $A_{21}^*(t, \delta, a, b, \gamma, \alpha)$ are rather cumbersome, which argues for imperfectness of the method used, but they are given in the explicit form in terms of elementary functions which allows a fast evaluation by use of a computer.

Remark 3.4.3. It can be made sure that in Theorem 3.4.1 we have: $A(t) \to 0$, $A_3(t) \to 1$, $A_7(t) \to 1$, $A_{16}(t) \to \infty$ as $t \to \infty$, so that all the conditions of Theorem 3.4.1, except (3.4.5) and (3.4.11), are trivially satisfied for $t \to \infty$ with every fixed a, b, γ. Moreover, for fixed a, b, γ, α, surely under assumption (3.4.11) (which is stronger than (3.4.5)), we have

$$\lim_{t\to\infty} A_{21}(t, \delta, a, b, \gamma, \alpha) = \gamma^{-(2+\delta)} \geqslant \lim_{\gamma\to1-} \gamma^{-(2+\delta)} = 1,$$

$$\lim_{t\to\infty} A_{21}^*(t, \delta, a, b, \gamma, \alpha) = 0.$$

Condition (3.4.11) is obviously fulfilled as $\gamma \to 1-$ for every fixed $b > 1$ and $0 < \delta \leqslant 1$.

3.4.5 The Main Result

For $s > 0$ let

$$Q_s(\delta) = \inf_{t, a, b, \gamma} \{C_s(\delta)t^{2+\delta} \vee A_{21}(t, \delta, a, b, \gamma, 0) \vee P(\delta, a, b)\}, \quad (3.4.26)$$

where the greatest lower bound is taken over the set of all possible values of the parameters t, a, b, γ satisfying conditions of Theorem 3.4.1, and

$$A_{21}^*(t, \delta, a, b, \gamma, 0) \leqslant s\, C_s(\delta)t^{2+\delta}, \quad t \geqslant 1/\sqrt{2\pi}, \quad a \leqslant P(\delta, a, b)/\varkappa$$

(the quantity $\varkappa = 0.54\ldots$ is defined in (3.2.6)). For $s = 0$ let

$$Q_0(\delta) = \inf_{t, a, b, \gamma} \{C_0(\delta)t^{2+\delta} \vee (A_{21}(t, \delta, a, b, \gamma, 1)$$
$$+ A_{21}^*(t, \delta, a, b, \gamma, 1)) \vee P(\delta, a, b)\}, \quad (3.4.27)$$

where the greatest lower bound is taken over the set of all possible values of the parameters t, a, b, γ satisfying conditions of Theorem 3.4.1, and $t \geqslant 1/\sqrt{2\pi}$, $a \leqslant P(\delta, a, b)/\varkappa$.

Theorem 3.4.4. *For all $s \geqslant 0$, $n \in \mathbb{N}$, and $F_1, F_2, \ldots, F_n \in \mathcal{F}_{2+\delta}$ we have*

$$\sup_{x \in \mathbb{R}} |x|^{2+\delta}\Delta_n(x) \leqslant \min_{0 \leqslant s \leqslant 1} Q_s(\delta)(\ell_n + s\tau_n).$$

Corollary 3.4.5. *Inequality* (3.1.11)

$$\sup_{x \in \mathbb{R}}(1 + |x|^{2+\delta})\Delta_n(x) \leqslant \min_{0 \leqslant s \leqslant 1} K_s(\delta)(\ell_n + s\tau_n)$$

holds for all $n \in \mathbb{N}$ and $F_1, F_2, \ldots, F_n \in \mathcal{F}_{2+\delta}$ with $K_s(\delta) = Q_s(\delta) + C_s(\delta)$.

Remark 3.4.6. For every $0 < \delta \leqslant 1$ the optimal values of the parameters t, a, b, γ satisfy

$$Q_s(\delta) = C_s(\delta)t^{2+\delta} = A_{21}(t, \delta, a, b, \gamma, 0) = P(\delta, a, b), \quad 0 < s \leqslant 1,$$
$$Q_0(\delta) = C_0(\delta)t^{2+\delta} = A_{21}(t, \delta, a, b, \gamma, 1) + A_{21}^*(t, \delta, a, b, \gamma, 1) = P(\delta, a, b).$$

Usually the value of $A_{21}^*(t_s, \delta, a_s, b_s, \gamma_s, 0)$ is considerably smaller than $sQ_s(\delta)$ (see Table 3.5).

Table 3.4 contains the values of the constants $Q_0(\delta)$, $K_0(\delta)$ for some $\delta \in (0, 1]$ in the non-i.i.d. case (an upper part), as well as in the i.i.d. case (a lower part). The optimal values $t_0(\delta)$, $a_0(\delta)$, $b_0(\delta)$, $\gamma_0(\delta)$ of the parameters t, a, b, γ, delivering minimum in (3.4.27) are also given in Table 3.4.

Table 3.5 contains the values of the constants $Q_s(\delta)$, $K_s(\delta)$ for some $\delta \in (0, 1]$ with $s = s_1(\delta)$ (this value of s minimizes the constant $C_s(\delta)$ in the uniform estimate). Also Table 3.5 contains the optimal values $t_s(\delta)$, $a_s(\delta)$, $b_s(\delta)$, $\gamma_s(\delta)$ of the parameters t, a, b, γ, that deliver minimum in (3.4.26) for the specified $s = s_1(\delta)$, as well as the value of $A_{21}^*(t, \delta, a, b, \gamma, 0)$ in the optimal point (with the specified t, a, b, γ). It is worth mentioning that the constants $K_s(\delta)$, $Q_s(\delta)$ do not satisfy the relation similar to $C_0(\delta) = (1 + s)C_s(\delta)$, which is true for the constants $C_s(\delta)$ from the uniform estimates, because the functions $A_{12}(x)$ and $A_{20}(x)$ contain some operations of minimum and maximum resulting to that the sums $A_{12}(x) + A_{12}^*(x)$ and $A_{20}(x) + A_{20}^*(x)$ for every $\alpha < 1$ are greater than the same sums with $\alpha = 1$.

The optimization algorithm in (3.4.26) and (3.4.27) has been realized for every $\delta \in [0, 1]$ and $s \geqslant 0$ in Matlab 7.12.0 by use of the procedure fminsearch(...) with the arguments t, a, b, γ.

Proof of Theorem 3.4.4. For the sake of brevity we shall omit the arguments of the functions $A_{21}(t, \delta, a, b, \gamma, \alpha)$ and $A_{21}^*(t, \delta, a, b, \gamma, \alpha)$. It suffices to prove that for an arbitrary set of the parameters t, a, b, γ satisfying the conditions of the theorem, for all $\alpha \in [0, 1]$, $s \geqslant 0$, $x \in \mathbb{R}$, $n \geqslant 1$, and $F_1, \ldots, F_n \in \mathcal{F}_{2+\delta}$ the inequality

$$|x|^{2+\delta} \Delta_n(x) \leqslant t^{2+\delta} C_s(\delta)(\ell_n + s\tau_n) \vee (A_{21}\ell_n + A_{21}^*\tau_n) \vee P(\delta, a, b)\ell_n$$

holds. Fix t and x. Then there are two possibilities:

1. $|x| \leqslant t$, then $|x|^{2+\delta} \Delta_n(x) \leqslant t^{2+\delta} C_s(\delta)(\ell_n + s\tau_n)$ for every $0 \leqslant s \leqslant 1$ by (3.1.6);
2. $|x| > t$, then for the given a, b, n, and $F_1, \ldots, F_n \in \mathcal{F}_{2+\delta}$ there are three possibilities:
   a. $c_n(x; \delta, a, b) \leqslant 0$, or, in an equivalent form, $a\ell_n \geqslant |x|^{2+\delta}$, whence, by (3.2.6), we conclude that $\Delta_n(x) \leqslant \varkappa \leqslant P(\delta, a, b)/a \leqslant P(\delta, a, b)\ell_n/|x|^{2+\delta}$;
   b. $0 < c_n(x; \delta, a, b) < x^2$, that is, $x^2 \geqslant (2\pi)^{-1} \vee c_n(x; \delta, a, b)$ and thus the conditions of Theorem 3.3.1 are met, according to which, $|x|^{2+\delta} \Delta_n(x) \leqslant P(\delta, a, b)\ell_n$;
   c. $x^2 \leqslant c_n(x; \delta, a, b)$. Taking also into account the condition $x^2 \geqslant t^2$, by Theorem 3.4.1 we obtain $|x|^{2+\delta} \Delta_n(x) \leqslant A_{21}\ell_n + A_{21}^*\tau_n$ for every $0 < \gamma < 1$ satisfying the conditions of the theorem and for arbitrary $\alpha \in [0, 1]$.

$\square$

Theorems 3.3.1–3.4.4 allow to describe an algorithm of evaluation of the functions $Q_s(t, \delta)$, $Q_s^*(t, \delta)$ for every $0 < \delta \leqslant 1$, that guarantee the validity of inequality (3.1.13):

$$\sup_{|x| \geqslant t} |x|^{2+\delta} \Delta_n(x) \leqslant \inf_{s \geqslant 0} \{Q_s(t, \delta)\ell_n + Q_s^*(t, \delta)\tau_n\} \leqslant \inf_{s \geqslant 0} Q_s(t, \delta)(\ell_n + s\tau_n),$$

$$t \geqslant 0,$$

for all $F_1, \ldots, F_n \in \mathcal{F}_{2+\delta}$ and $n \geqslant 1$.

Theorem 3.4.7. *For all* $0 < \delta \leqslant 1$, $n \in \mathbb{N}$, $F_1, \ldots, F_n \in \mathcal{F}_{2+\delta}$, *and every* $t \geqslant 0$ *inequality* (3.1.13) *holds with*

$$Q_s(t, \delta) := \begin{cases} Q_s(\delta), & t < t_s(\delta), \\ \inf_{a, b, \gamma} \max\{A_{21}(t, \delta, a, b, \gamma, 0),\ P(\delta, a, b)\}, & t \geqslant t_s(\delta), \end{cases}$$

$$Q_s^*(t, \delta) := \begin{cases} sQ_s(\delta), & t < t_s(\delta), \\ A_{21}^*(t, \delta, a(t), b(t), \gamma(t), 0), & t \geqslant t_s(\delta), \end{cases} \quad \text{for } s \in (0, 1],$$

where $t_s(\delta) \geqslant 1/\sqrt{2\pi}$ *is the optimal value of the parameter* t *in* (3.4.26); *the greatest lower bound is taken for every* $t \geqslant t_s(\delta)$ *over the set all possible values of the parameters* a, b, γ, *satisfying the conditions of Theorem 3.4.1, and the conditions*

$$A_{21}^*(t, \delta, a, b, \gamma, 0) \leqslant sA_{21}(t, \delta, a, b, \gamma, 0), \quad a \leqslant P(\delta, a, b)/\varkappa$$

(the quantity $\varkappa = 0.54\ldots$ *is defined in* (3.2.6)); $a(t), b(t), \gamma(t)$ *being the values of the parameters* a, b, γ, *that deliver minimum in the definition of* $Q_s(t, \delta)$.

For $s = 0$ *inequality* (3.1.13) *holds with* $Q_0^*(t, \delta) = 0$ *and*

$$Q_0(t, \delta) = \begin{cases} Q_0(\delta), & t < t_0(\delta), \\ \inf_{a, b, \gamma} \{(A_{21}(t, \delta, a, b, \gamma, 1) + A_{21}^*(t, \delta, a, b, \gamma, 1)) \vee P(\delta, a, b)\}, & t \geqslant t_0(\delta), \end{cases}$$

where $t_0(\delta) \geqslant 1/\sqrt{2\pi}$ *is the optimal value of the parameter* t *in* (3.4.27), *and for* $t \geqslant t_0(\delta)$ *the greatest lower bound for is taken over the set of all possible values of the parameters* a, b, γ, *satisfying the conditions of Theorem 3.4.1, as well as the condition* $a \leqslant P(\delta, a, b)/\varkappa$.

Moreover, for every $s \in [0, 1]$ *and* $0 < \delta \leqslant 1$

$$\lim_{t \to \infty} Q_s(t, \delta) = 1 + e = 3.7182\ldots, \quad \lim_{t \to \infty} Q_s^*(t, \delta) = 0.$$

Corollary 3.4.8. *For all* $0 < \delta \leqslant 1$, $F_1, \ldots, F_n \in \mathcal{F}_{2+\delta}$, $n \in \mathbb{N}$, *and every* $t > 0$

$$\sup_{|x| \geqslant t} (1 + |x|^{2+\delta}) \Delta_n(x)$$

$$\leqslant \inf_{s \geqslant 0} \left\{ Q_s(t, \delta) \ell_n + Q_s^*(t, \delta) \tau_n + \min \left\{ \frac{Q_s(t, \delta) \ell_n + Q_s^*(t, \delta) \tau_n}{t^{2+\delta}}, \right. \right.$$

$$\left. \left. \min_{0 \leqslant q \leqslant 1} C_q(\ell_n + q\tau_n) \right\} \right\}.$$

Proof of Theorem 3.4.7. For the sake of brevity we shall omit the arguments of the functions $A_{21}(t, \delta, a, b, \gamma, \alpha)$ and $A_{21}^*(t, \delta, a, b, \gamma, \alpha)$. It is easy to see that for all $s, t \geqslant 0$

$$\sup_{|x| \geqslant t} |x|^{2+\delta} \Delta_n(x) \leqslant \sup_{x \in \mathbb{R}} |x|^{2+\delta} \Delta_n(x) \leqslant Q_s(\delta) \ell_n + s Q_s(\delta) \tau_n$$

by Theorem 3.4.4. Thus, it suffices to prove that for arbitrary $t \geqslant t_s(\delta)$, $s \geqslant 0$ and every $n \in \mathbb{N}$, $F_1, \ldots, F_n \in \mathcal{F}_{2+\delta}$, $|x| \geqslant t$, $\alpha \in [0, 1]$, and a, b, γ, satisfying the conditions of the theorem, the inequality

$$|x|^{2+\delta} \Delta_n(x) \leqslant (A_{21} \ell_n + A_{21}^* \tau_n) \vee P(\delta, a, b) \ell_n.$$

holds. Fixing t, a, b yields two possibilities:

1. $c_n(t; \delta, a, b) \leqslant 0$. Then for $|x| \geqslant t$ there are three possibilities:
   (a) $c_n(x; \delta, a, b) \leqslant 0$, or, in an equivalent form, $a\ell_n/|x|^{2+\delta} \geqslant 1$, whence by (3.2.6) we conclude that $\Delta_n(x) \leqslant \varkappa \leqslant P(\delta, a, b)/a \leqslant P(\delta, a, b) \ell_n/|x|^{2+\delta}$.
   (b) $0 < c_n(x; \delta, a, b) \leqslant x^2$. Since we also have $x^2 \geqslant t^2 \geqslant (2\pi)^{-1}$, by Theorem 3.3.1 we conclude that $|x|^{2+\delta} \Delta_n(x) \leqslant P(\delta, a, b) \ell_n$.
   (c) $c_n(x; \delta, a, b) \geqslant x^2 \geqslant t^2$. In this case, by Theorem 3.4.1, for all $\alpha \in [0, 1]$ we have

$$|x|^{2+\delta} \Delta_n(x) \leqslant A_{21} \ell_n + A_{21}^* \tau_n.$$

2. $c_n(t; \delta, a, b) > 0$. Since $c_n(x; \delta, a, b) > c_n(t; \delta, a, b) > 0$ for all $|x| \geqslant t$, there are two possibilities for $c_n(x; \delta, a, b)$:
   (a) $0 < c_n(x; \delta, a, b) \leqslant x^2$. Since we also have $x^2 \geqslant t^2 \geqslant (2\pi)^{-1}$, by Theorem 3.3.1 we conclude that $|x|^{2+\delta} \Delta_n(x) \leqslant P(\delta, a, b) \ell_n$.
   (b) $c_n(x; \delta, a, b) \geqslant x^2 \geqslant t^2$, then Theorem 3.4.1 yields

$$|x|^{2+\delta} \Delta_n(x) \leqslant A_{21} \ell_n + A_{21}^* \tau_n.$$

Let us find the limiting values of the functions $Q_s(t, \delta)$, $Q_s^*(t, \delta)$ as $t \to \infty$. On one hand, for every $0 < \delta \leqslant 1$, $s \geqslant 0$, and $t \geqslant 0$ we have

$$Q_s(t, \delta) \geqslant \inf_{a>0, b>1} P(\delta, a, b) = 1 + e.$$

On the other hand, for fixed $\delta, a, b, \gamma, \alpha$ by Remark 3.4.3 we have

$$\lim_{t \to \infty} A_{21}(t, \delta, a, b, \gamma, \alpha) = \gamma^{-(2+\delta)}, \quad \lim_{t \to \infty} A_{21}^*(t, \delta, a, b, \gamma, \alpha) = 0,$$

and hence, for every $s \in [0, 1]$

$$\lim_{t \to \infty} Q_s(t, \delta) \leqslant \lim_{t \to \infty} Q_0(t, \delta)$$
$$\leqslant \inf_{a, b, \gamma} \lim_{t \to \infty} (A_{21}(t, \delta, a, b, \gamma, 1) + A_{21}^*(t, \delta, a, b, \gamma, 1)) \vee$$
$$P(\delta, a, b)$$
$$= \inf_{a, b, \gamma} \max \left\{ b^{2+\delta} + a \exp \left\{ b^{2+\delta}/a \right\}, \gamma^{-(2+\delta)} \right\} = 1 + e,$$

where the greatest lower bound is taken over the set of all $a > 0$, $b > 1$, $\gamma \in (0, 1)$ satisfying the conditions $a \leqslant P(\delta, a, b)/\varkappa$ and (3.4.11), and is attained at $a = b^{2+\delta}$ (such a choice of a is possible for every $b > 1$, because $\varkappa < 1 + e$) and letting, first, $\gamma \to 1-$, and then $b \to 1+$.   □

Remark 3.4.9. Since in the definition of $Q_s(t, \delta)$ with $t > t_s(\delta)$, $s > 0$, the maximum of A_{21} and P is minimized, the optimal value is attained, when $A_{21}(t, \delta, a, b, \gamma, 0) = P(\delta, a, b)$, and hence, $Q_s(t, \delta) = P(\delta, a(t), b(t)) = A_{21}(t, \delta, a(t), b(t), \gamma(t), 0) \geqslant A_{21}^*(t, \delta, a(t), b(t), \gamma(t), 0)/s = Q_s^*(t, \delta)/s$ for every $s > 0$ by the conditions of minimization. Thus,

$$Q_s^*(t, \delta) \leqslant s Q_s(t, \delta), \quad s \geqslant 0, \ t \geqslant 0.$$

Remark 3.4.10. Generally speaking, the function $Q_s^*(t, \delta)$ has a discontinuity at $t = t_s(\delta)$ for every $s > 0$, since

$$\lim_{t \to t_s(\delta)-} Q_s^*(t, \delta) = s Q_s(\delta) = s A_{21}(t_s, \delta, a_s, b_s, \gamma_s, 0)$$
$$\geqslant A_{21}^*(t_s, \delta, a_s, b_s, \gamma_s, 0),$$

where $t_s = t_s(\delta)$, $a_s = a(t_s)$, $b_s = b(t_s)$, $\gamma_s = \gamma(t_s)$, and the values of $A_{21}^*(t_s, \delta, a_s, b_s, \gamma_s, 0)$ are usually considerably smaller than $s Q_s(\delta)$ for $s > 0$ under consideration (see, e.g., Table 3.5).

The values of the functions $Q_s(t, \delta)$ for some $\delta \in (0, 1]$ and $t \geqslant t_s(\delta)$, $t = 0$ are given in Table 3.6 for $s = 0$ and in Table 3.7 (in the third and

Table 3.6 Upper Bounds for $Q_0(t, \delta)$ From Theorem 3.4.7 ($Q_0^*(t, \delta) \equiv 0$) for Some $t \geqslant 0$ and $\delta \in (0, 1]$

$t \backslash \delta$	1.0	0.9	0.8	0.7	0.6	0.5	0.4	0.3	0.2	0.1
Non-i.i.d. Case With $s = 0$										
0.0	21.26	19.51	17.95	16.55	15.30	14.19	13.23	12.35	11.54	10.79
3.6	19.78	18.33	16.98	15.73	14.57	13.49	12.48	11.55	10.68	9.88
3.7	19.17	17.77	16.46	15.26	14.13	13.09	12.12	11.21	10.38	9.60
3.8	18.57	17.19	15.91	14.73	13.65	12.65	11.73	10.88	10.08	9.34
3.9	17.87	16.54	15.32	14.20	13.17	12.22	11.34	10.53	9.79	9.08
4.0	17.19	15.93	14.77	13.70	12.72	11.81	10.98	10.21	9.49	8.84
5.0	12.35	11.55	10.81	10.13	9.49	8.90	8.36	7.86	7.41	6.99
6.0	9.60	9.05	8.54	8.07	7.65	7.26	6.92	6.62	6.32	6.07
7.0	8.44	7.91	7.44	7.03	6.69	6.42	6.22	6.04	5.88	5.72
8.0	8.02	7.55	7.13	6.75	6.43	6.15	5.94	5.80	5.67	5.55
9.0	7.67	7.24	6.86	6.52	6.22	5.96	5.74	5.60	5.49	5.39
10.0	7.36	6.98	6.63	6.33	6.05	5.81	5.59	5.44	5.34	5.25
50.0	4.56	4.50	4.44	4.38	4.33	4.28	4.23	4.18	4.14	4.10
∞	3.72	3.72	3.72	3.72	3.72	3.72	3.72	3.72	3.72	3.72
I.i.d. Case With $s = 0$										
0.0	16.90	15.75	14.68	13.61	12.62	11.72	10.88	10.09	9.34	8.60
3.6	15.85	14.80	13.84	12.95	12.11	11.32	10.58	9.87	9.21	8.58
3.7	15.54	14.52	13.57	12.69	11.86	11.08	10.35	9.66	9.01	8.40
3.8	15.22	14.22	13.29	12.42	11.61	10.84	10.12	9.45	8.81	8.21
3.9	14.90	13.92	13.01	12.15	11.36	10.60	9.90	9.24	8.61	8.03
4.0	14.58	13.61	12.72	11.89	11.10	10.37	9.68	9.03	8.42	7.86
5.0	11.56	10.81	10.11	9.46	8.85	8.28	7.76	7.27	6.83	6.43
6.0	9.22	8.66	8.13	7.65	7.21	6.81	6.45	6.12	5.83	5.60
7.0	7.52	7.12	6.77	6.46	6.18	5.94	5.77	5.65	5.54	5.44
8.0	6.47	6.24	6.05	5.92	5.80	5.69	5.58	5.48	5.38	5.28
9.0	6.05	5.93	5.82	5.72	5.62	5.52	5.43	5.33	5.24	5.16
10.0	5.85	5.75	5.65	5.56	5.47	5.38	5.30	5.22	5.13	5.05
50.0	4.25	4.23	4.22	4.20	4.18	4.16	4.14	4.12	4.10	4.09
∞	3.72	3.72	3.72	3.72	3.72	3.72	3.72	3.72	3.72	3.72

subsequent columns) for $s = s_1(\delta)$. For $Q_s^*(t, \delta)$ with $s > 0$ the following upper bounds can be used

$$Q_s^*(t, \delta) \leqslant \begin{cases} sQ_s(t, \delta) \leqslant sQ_s(\delta), & \forall t \geqslant 0, \\ A_{21}^*(t, \delta, a(t), b(t), \gamma(t), 0) \leqslant A_{21}^*(t_s, \delta, a_s, b_s, \gamma_s, 0), & t \geqslant t_s(\delta). \end{cases}$$

Table 3.7 Upper Bounds for $Q_s^*(t, 1)$ (the Second Column) and $Q_s(t, \delta)$ (the Third and Subsequent Columns) From Theorem 3.4.7 for Some $t \geqslant t_s(\delta)$ and $\delta \in (0, 1]$ With $s = s_1(\delta)$ Which Is the Minimal Point of Minimum of $C_s(\delta)$ (Values of $s_1(\delta)$ Are Given in Table 3.5)

$t \backslash \delta$	1.0		0.9	0.8	0.7	0.6	0.5	0.4	0.3	0.2	0.1
Non-i.i.d. Case With $s = s_1(\delta)$											
$t_s(\delta)$	0.210	17.88	16.34	15.02	13.86	12.83	11.92	11.12	10.41	9.79	9.22
3.9	0.199	17.76	16.34	15.02	13.86	12.83	11.92	11.12	10.41	9.78	9.08
4.0	0.150	17.10	15.87	14.73	13.68	12.71	11.81	10.97	10.20	9.49	8.83
5.0	0.006	12.34	11.55	10.81	10.12	9.49	8.90	8.36	7.86	7.41	6.99
6.0	0.001	9.60	9.05	8.54	8.07	7.65	7.26	6.92	6.62	6.32	6.07
7.0	0.001	8.47	7.91	7.44	7.03	6.69	6.42	6.22	6.04	5.88	5.72
8.0	0.001	8.02	7.55	7.13	6.75	6.43	6.15	5.94	5.80	5.67	5.55
9.0	0.001	7.67	7.24	6.86	6.52	6.22	5.96	5.74	5.60	5.49	5.39
10.0	0.001	7.36	6.98	6.63	6.33	6.05	5.81	5.59	5.44	5.34	5.25
50.0	0.001	4.56	4.50	4.44	4.38	4.33	4.28	4.23	4.18	4.14	4.10
∞	0	3.72	3.72	3.72	3.72	3.72	3.72	3.72	3.72	3.72	3.72
I.i.d. Case With $s = s_1(\delta)$											
$t_s(\delta)$	0.166	15.40	14.30	13.29	12.38	11.55	10.79	10.11	9.49	8.93	8.43
3.9	0.100	14.82	13.88	12.99	12.15	11.35	10.60	9.90	9.24	8.61	8.03
4.0	0.077	14.51	13.58	12.71	11.88	11.10	10.37	9.68	9.03	8.42	7.86
5.0	0.005	11.55	10.81	10.11	9.46	8.85	8.28	7.76	7.27	6.83	6.43
6.0	0.001	9.22	8.66	8.13	7.65	7.21	6.81	6.45	6.12	5.83	5.60
7.0	0.001	7.52	7.12	6.77	6.46	6.18	5.94	5.77	5.65	5.54	5.44
8.0	0.001	6.47	6.24	6.05	5.92	5.80	5.69	5.58	5.48	5.38	5.28
9.0	0.001	6.05	5.93	5.82	5.72	5.62	5.52	5.43	5.33	5.24	5.16
10.0	0.001	5.85	5.75	5.65	5.56	5.47	5.38	5.30	5.22	5.13	5.05
50.0	0.001	4.25	4.23	4.22	4.20	4.18	4.16	4.14	4.12	4.10	4.09
∞	0	3.72	3.72	3.72	3.72	3.72	3.72	3.72	3.72	3.72	3.72

Values of $s = s_1(\delta)$, $Q_s(\delta)$, and $A_{21}^*(t_s, \delta, a_s, b_s, \gamma_s, 0)$ are given in the sixth, eighth, and seventh columns of Table 3.5, respectively. As it can be seen from Table 3.7, with the account of $Q_s(t, \delta) \geqslant 1 + e$, $t \geqslant 0$, for all $t \geqslant t_s(\delta)$ under consideration the values of $Q_s^*(t, \delta)$ are substantially smaller than the values of $sQ_s(t, \delta)$ with $s = s_1(\delta)$.

In view of the great importance of the case $\delta = 1$ we also evaluate $Q_s^*(t, 1) = A_{21}^*(t, 1, a(t), b(t), \gamma(t), 0)$ for $s = s_1(1)$ and some $t \geqslant t_s(1)$ (see the second column in Table 3.7). In particular, we have

Corollary 3.4.11. *For all $F_1, \ldots, F_n \in \mathcal{F}_3$ and $n \in \mathbb{N}$*

$$\sup_{|x| \geqslant t} |x|^3 \Delta_n(x) \leqslant \begin{cases} 17.88(\ell_n + \tau_n), & t \geqslant 0, \\ 17.88\ell_n + 0.21\tau_n, & t \geqslant 3.89, \\ 17.10\ell_n + 0.15\tau_n, & t \geqslant 4, \\ 12.34\ell_n + 0.006\tau_n, & t \geqslant 5. \end{cases}$$

For every $F_1 = \cdots = F_n \in \mathcal{F}_3$ and $n \in \mathbb{N}$

$$\sup_{|x| \geqslant t} |x|^3 \Delta_n(x) \leqslant \begin{cases} 15.40(\ell_n + 0.646\tau_n), & t \geqslant 0, \\ 15.40\ell_n + 0.166\tau_n, & t \geqslant 3.71, \\ 14.51\ell_n + 0.077\tau_n, & t \geqslant 4, \\ 11.55\ell_n + 0.005\tau_n, & t \geqslant 5. \end{cases}$$

Theorem 3.4.7 also yields upper bounds for the Kolmogorov function

$$D^*(x, \delta) = \sup_{n \geqslant 1, \, F_1, \ldots, F_n \in \mathcal{F}_{2+\delta}} \Delta_n(x)/\ell_n, \quad x \in \mathbb{R}.$$

Corollary 3.4.12. *For every $0 < \delta \leqslant 1$ we have*

$$\sup_{|x| \geqslant t} |x|^{2+\delta} D^*(x, \delta) \leqslant Q_0(t, \delta), \quad t \geqslant 0,$$

$$1 \leqslant \limsup_{|x| \to \infty} |x|^{2+\delta} D^*(x, \delta) \leqslant 1 + e < 3.7183,$$

$$\sup_{n \geqslant 1, \, F_1, \ldots, F_n \in \mathcal{F}_{2+\delta}} \limsup_{|x| \to \infty} |x|^{2+\delta} \Delta_n(x)/\ell_n \leqslant 1.$$

Proof. The first two upper bounds for $D^*(x, \delta)$ follow from the definition and the properties of the function $Q_0(t, \delta)$. The lower bound for $D^*(x, \delta)$ follows from [9] (see (3.1.12)). To prove the third upper bound it suffices to note that for fixed $n \in \mathbb{N}$ and $F_1, \ldots, F_n \in \mathcal{F}_{2+\delta}$ the value of the Lyapunov fraction ℓ_n also remains fixed (separated from zero and from infinity), and hence, for arbitrary fixed $a > 0$, $b > c > 1$ the conditions

$$x^2 \geqslant \frac{b^2}{2(b-c)} \ln \frac{|x|^{2+\delta}}{a\ell_n} \geqslant \frac{1}{2\pi},$$

which guarantee the validity of Theorem 3.3.1, are fulfilled for all sufficiently large $|x|$. Thus, by Theorem 3.3.1, for every $a > 0$ and $b > 1$ with $c = (b+1)/2 \in (1, b)$ we have

$$\limsup_{|x|\to\infty} |x|^{2+\delta}\Delta_n(x)/\ell_n$$

$$\leqslant b^{2+\delta} + a\exp\left\{\frac{b^{2+\delta}}{a}\right\}$$

$$\times \limsup_{|x|\to\infty}\left(\frac{a\ell_n}{x^{2+\delta}}\right)^{(b-1)/2} = b^{2+\delta}.$$

Now letting $b \to 1+$, we obtain the claim. $\square$

Acknowledgments

The work is supported by the Russian Foundation for Basic Research (projects 15-07-02984-a, 16-31-60110-mol-a-dk) and by the Ministry for Education and Science of Russia (grant No. MD-5642.2015.1).

REFERENCES

[1] I.A. Ahmad, P.-E. Lin, A Berry–Esseen type theorem, Utilitas Math. 11 (1977) 153–160.

[2] A.D. Barbour, P. Hall, Stein's method and the Berry–Esseen theorem, Aust. J. Stat. 26 (1984) 8–15.

[3] A.C. Berry, The accuracy of the Gaussian approximation to the sum of independent variates, Trans. Am. Math. Soc. 49 (1941) 122–136.

[4] R.N. Bhattacharya, R. Ranga Rao, Normal Approximation and Asymptotic Expansions, Wiley, New York, 1976.

[5] A. Bikelis, Estimates of the remainder term in the central limit theorem, Litovsk. Mat. Sb. 6 (3) (1966) 323–346 (in Russian).

[6] A. Bikelis, On the accuracy of the approximation of distributions of sums of independent identically distributed random variables by the normal distribution, Litovsk. Mat. Sb. 11 (2) (1971) 237–240 (in Russian).

[7] S.G. Bobkov, Closeness of probability distributions in terms of Fourier–Stieltjes transforms, Rus. Math. Surv. 71 (6) (2016) 1021–1079.

[8] L.H.Y. Chen, Q.M. Shao, A non-uniform Berry–Esseen bound via Stein's method, Probab. Theory Relat. Fields 120 (2001) 236–254.

[9] G.P. Chistyakov, On a problem of A.N. Kolmogorov, J. Math. Sci. 68 (4) (1994) 604–625.

[10] H. Cramér, Sur un nouveau théorèm-limite de la théorie des probabilités, Actualités Sci. Indust. 736 (1938) 5–23.

[11] C.-G. Esseen, On the Liapounoff limit of error in the theory of probability, Ark. Mat. Astron. Fys. A28 (9) (1942) 1–19.

[12] C.-G. Esseen, Fourier analysis of distribution functions. A mathematical study of the Laplace-Gaussian law, Acta Math. 77 (1) (1945) 1–125.

[13] C.-G. Esseen, A moment inequality with an application to the central limit theorem, Skand. Aktuarietidskr. 39 (1956) 160–170.

[14] W. Feller, On the Berry–Esseen theorem, Z. Wahrsch. Verw. Geb. 10 (1968) 261–268.

[15] S.V. Gavrilenko, An improvement of the nonuniform estimates of convergence rate of distributions of Poisson random sums to the normal law, Inform. Appl. 5 (1) (2011) 12–24 (in Russian).

[16] M.E. Grigorieva, S.V. Popov, An upper bound for the absolute constant in the nonuniform version of the Berry–Esseen inequalities for nonidentically distributed summands, Dokl. Math. 86 (1) (2012) 524–526.

[17] M.E. Grigorieva, S.V. Popov, On nonuniform convergence rate estimates in the central limit theorem, Syst. Means Inform. 22 (1) (2012) 180–204 (in Russian).

[18] C.C. Heyde, A nonuniform bound on convergence to normality, Ann. Probab. 3 (5) (1975) 903–907.

[19] M.L. Katz, Note on the Berry–Esseen theorem, Ann. Math. Stat. 34 (1963) 1107–1108.

[20] A.N. Kolmogorov, Some recent works in the field of limit theorems of probability theory, Bull. Moscow Univ. 10 (7) (1953) 29–38 (in Russian).

[21] V. Korolev, I. Shevtsova, An improvement of the Berry–Esseen inequality with applications to Poisson and mixed Poisson random sums, Scand. Actuar. J. 2012 (2) (2012) 81–105.

[22] V.Yu. Korolev, A.V. Dorofeyeva, Bounds of the accuracy of the normal approximation to the distributions of random sums under relaxed moment conditions, Lith. Math. J. (2017) (in press).

[23] V.Yu. Korolev, S.V. Popov, Improvement of convergence rate estimates in the central limit theorem under the absence of moments of orders greater than the second, Theory Probab. Appl. 56 (4) (2012) 682–691.

[24] V.Yu. Korolev, S.V. Popov, Improvement of convergence rate estimates in the central limit theorem under weakened moment conditions, Dokl. Math. 86 (1) (2012) 506–511.

[25] V.Yu. Korolev, I.G. Shevtsova, On the upper bound for the absolute constant in the Berry–Esseen inequality, Theory Probab. Appl. 54 (4) (2010) 638–658.

[26] V.Yu. Korolev, I.G. Shevtsova, A new moment-type estimate of convergence rate in the Lyapunov theorem, Theory Probab. Appl. 55 (3) (2011) 505–509.

[27] W.Y. Loh, On the normal approximation for sums of mixing random variables, Master thesis, Department of Mathematics, University of Singapore, 1975.

[28] M. Maejima, A note on the nonuniform rate of convergence to normality, Yokohama Math. J. 28 (1–2) (1980) 97–106.

[29] L.D. Meshalkin, B.A. Rogozin, An estimate of the distance between distribution functions by the closeness of their characteristic functions and its application to the central limit theorem, in: Limit Theorems of Probability, Uzbekistan Academy of Sciences Publishing, Tashkent, USSR, 1963, pp. 49–55 (in Russian).

[30] R. Michel, On the accuracy of nonuniform Gaussian approximation to the distribution functions of sums of independent and identically distributed random variables, Z. Wahrsch. Verw. Geb. 35 (4) (1976) 337–347.

[31] R. Michel, On the constant in the nonuniform version of the Berry–Esseen theorem, Z. Wahrsch. Verw. Geb. 55 (1) (1981) 109–117.

[32] Sh.A. Mirachmedov, On the absolute constant in the nonuniform convergence rate estimate in the central limit theorem, Izv. AN UzSSR, Ser. Fiz.-Mat. Nauk 4 (1984) 26–30 (in Russian).

[33] S.V. Nagaev, Some limit theorems for large deviations, Theory Probab. Appl. 10 (2) (1965) 214–235.

[34] K. Neammanee, On the constant in the nonuniform version of the Berry–Esseen theorem, Int. J. Math. Math. Sci. 12 (2005) 1951–1967.

[35] K. Neammanee, P. Thongtha, Improvement of the non-uniform version of Berry–Esseen inequality via Paditz–Siganov theorems, J. Inequal. Pure Appl. Math. 8 (4) (2007).

[36] Yu.S. Nefedova, I.G. Shevtsova, On the accuracy of the normal approximation to distributions of Poisson random sums, Inform. Appl. 5 (1) (2011) 39–45 (in Russian).

[37] Yu.S. Nefedova, I.G. Shevtsova, Structural improvement of nonuniform estimates for the rate of convergence in the central limit theorem with applications to Poisson random sums, Dokl. Math. 84 (2) (2011) 675–680.

[38] Yu.S. Nefedova, I.G. Shevtsova, On non-uniform convergence rate estimates in the central limit theorem, Theory Probab. Appl. 57 (1) (2013) 28–59.

[39] V.N. Nikulin, On finite deviations, in: 5th Vilnius Conference on Probability Theory and Mathematical Statistics. Abstracts of Communications, Vilnius, June 26–July 1 1989, vol. 4, 1989, pp. 105–106.

[40] V.N. Nikulin, Nonuniform estimates of the remainder term in the central limit theorem, Theory Probab. Appl. 36 (4) (1992) 831–832.

[41] V.N. Nikulin, An algorithm to estimate a nonuniform convergence bound in the central limit theorem, 2010, arXiv:1004.0552.

[42] V.N. Nikulin, L. Paditz, A note on nonuniform CLT-bounds, in: 7th Vilnius Conference on Probability Theory and 22nd European Meeting of Statisticians. Abstracts, Vilnius, 1998, pp. 358–359.

[43] L.V. Osipov, Refinement of Lindeberg's theorem, Theory Probab. Appl. 11 (2) (1966) 299–302.

[44] L.V. Osipov, V.V. Petrov, On an estimate of the remainder term in the central limit theorem, Theory Probab. Appl. 12 (2) (1967) 281–286.

[45] L. Paditz, Abschätzungen der Konvergenzgeschwindigkeit im zentralen Grenzwertsatz, Wiss. Z. der TU Dresden 25 (1976) 1169–1177.

[46] L. Paditz, Über die Annäherung der Verteilungsfunktionen von Summen unabhängiger Zufallsgrößen gegen unbegrenzt teilbare Verteilungsfunktionen unter besonderer Beachtung der Verteilungsfunktion der standardisierten Normalverteilung, Dissertation A, Technische Universität Dresden, Dresden, 1977.

[47] L. Paditz, Abschätzungen der Konvergenzgeschwindigkeit zur Normalverteilung unter Voraussetzung einseitiger Momente, Math. Nachr. 82 (1978) 131–156.

[48] L. Paditz, Über eine Fehlerabschätzung im zentralen Grenzwertsatz, Wiss. Z. der TU Dresden 28 (5) (1979) 1197–1200.

[49] L. Paditz, Bemerkungen zu einer Fehlerabschätzung im zentralen Grenzwertsatz, Wiss. Z. Hochschule für Verkehrswesen, "Friedrich List". Dresden 27 (4) (1980) 829–837.

[50] L. Paditz, On error-estimates in the central limit theorem for generalized linear discounting, Math. Operationsforsch. u. Stat., Ser. Stat. 15 (4) (1984) 601–610.

[51] L. Paditz, Über eine globale Fehlerabschätzung im zentralen Grenzwertsatz, Wiss. Z. Hochschule für Verkehrswesen "Friedrich List". Dresden. 33 (2) (1986) 399–404.

[52] L. Paditz, On the analytical structure of the constant in the nonuniform version of the Esseen inequality, Statistics 20 (3) (1989) 453–464.

[53] L. Paditz, Sh. A. Mirachmedov, Pis'mo v redaciju (Zamechanie k ocenke absolutnoj postojannoj v neravnomernoj ocenke skorosti shodimosti v c.p.t.), Izv. AN UzSSR, Ser. Fiz.-Mat. Nauk 3 (1986) 80 (in Russian).

[54] V.V. Petrov, An estimate of the deviation of the distribution function of a sum of independent random variables from the normal law, Sov. Math. Dokl. 6 (5) (1965) 242–244.

[55] V.V. Petrov, Sums of Independent Random Variables, Springer-Verlag, Berlin/Heidelberg, 1975.

[56] V.V. Petrov, A limit theorem for sums of independent, nonidentically distributed random variables, J. Sov. Math. 20 (3) (1982) 2232–2235.

[57] V.V. Petrov, Limit Theorems of Probability Theory. Sequences of Independent Random Variables, Clarendon Press, Oxford, 1995.

[58] V.V. Petrov, On estimation of the remainder in the central limit theorem, J. Math. Sci. 147 (4) (2007) 6929–6931.

[59] I. Pinelis, On the nonuniform Berry–Esseen bound, 2013, arXiv:1301.2828.

[60] I. Pinelis, Optimal re-centering bounds, with applications to Rosenthal-type concentration of measure inequalities, in: High Dimensional Probability VI, Progress in Probability, vol. 66, Springer, Basel, 2013, pp. 81–93.

[61] H. Prawitz, Limits for a distribution, if the characteristic function is given in a finite domain, Skand. Aktuarietidskr. 55 (1972) 138–154.

[62] H. Prawitz, On the remainder in the central limit theorem. I: One-dimensional independent variables with finite absolute moments of third order, Scand. Actuar. J. 3 (1975) 145–156.

[63] L.V. Rozovsky, A nonuniform estimate of the remainder in the central limit theorem, J. Math. Sci. 152 (6) (2008) 932–933.

[64] Z. Rychlik, Non-uniform central limit bounds with applications to probabilities of deviations, Theory Probab. Appl. 28 (4) (1984) 681–687.

[65] I. Shevtsova, On the absolute constants in the Berry–Esseen type inequalities for identically distributed summands, 2011, arXiv:1111.6554.

[66] I. Shevtsova, Moment-type estimates with asymptotically optimal structure for the accuracy of the normal approximation, Ann. Math. Inform. 39 (2012) 241–307.

[67] I.G. Shevtsova, Sharpening of the upper bound for the absolute constant in the Berry–Esseen inequality, Theory Probab. Appl. 51 (3) (2007) 549–553.

[68] I.G. Shevtsova, An improvement of convergence rate estimates in Lyapunov's theorem, Dokl. Math. 82 (3) (2010) 862–864.

[69] I.G. Shevtsova, On the asymptotically exact constants in the Berry–Esseen–Katz inequality, Theory Probab. Appl. 55 (2) (2011) 225–252.

[70] I.G. Shevtsova, On the absolute constant in the Berry–Esseen inequality and its structural and non-uniform improvements, Inform. Appl. 7 (1) (2013) 124–125 (in Russian).

[71] I.G. Shevtsova, On the absolute constants in the Berry–Esseen-type inequalities, Dokl. Math. 89 (3) (2014) 378–381.

[72] I.G. Shevtsova, A moment inequality with application to estimates of the rate of convergence in the global CLT for Poisson-binomial random sums, Theory Probab. Appl. 62 (2) (2017) (in Russian) in press.

[73] P. Thongtha, K. Neammanee, Refinement on the constants in the non-uniform version of the Berry–Esseen theorem, Thai J. Math. 5 (2007) 1–13.

[74] W. Tysiak, Gleichmäige und nicht-gleichmäige Berry–Esseen Abschätzungen, Dissertation, Gesamthochschule Wuppertal, Wuppertal, 1983.

[75] I.S. Tyurin, On the accuracy of the Gaussian approximation, Dokl. Math. 80 (3) (2009) 840–843.

[76] I.S. Tyurin, Refinement of the upper bounds of the constants in Lyapunov's theorem, Rus. Math. Surv. 65 (3) (2010) 586–588.

[77] I.S. Tyurin, On the convergence rate in Lyapunov's theorem, Theory Probab. Appl. 55 (2) (2011) 253–270.

[78] I.S. Tyurin, A refinement of the remainder in the Lyapunov theorem, Theory Probab. Appl. 56 (4) (2012) 693–696.

[79] I.S. Tyurin, Some optimal bounds in CLT using zero biasing, Stat. Prob. Lett. 82 (3) (2012) 514–518.

CHAPTER 4

On the Nonuniform Berry–Esseen Bound

Iosif Pinelis
Michigan Technological University, Houghton, MI, United States

4.1 UNIFORM AND NONUNIFORM BERRY–ESSEEN (BE) BOUNDS

Suppose that $X_1, \ldots, X_n$ are independent zero-mean random variables (r.v.'s), with

$$S := X_1 + \cdots + X_n, \quad A := \sum \mathsf{E}\,|X_i|^3 < \infty, \quad \text{and} \quad B := \sqrt{\sum \mathsf{E} X_i^2} > 0.$$

Consider

$$\Delta(z) := |\mathsf{P}(S > Bz) - \mathsf{P}(Z > z)| \quad \text{and} \quad r_L := A/B^3,$$

where $Z \sim N(0, 1)$ and $z \geqslant 0$; of course, r_L is the so-called Lyapunov ratio. Note that, in the "iid" case (when the X_i's are iid), r_L will be on the order of $1/\sqrt{n}$.

In such an iid case, let us also assume that $\mathsf{E} X_1^2 = 1$.

Uniform and nonuniform BE bounds are upper bounds on $\Delta(z)$ of the forms

$$c_{\mathsf{u}}\, r_L \quad \text{and} \quad c_{\mathsf{nu}} \frac{r_L}{1 + z^3}, \tag{4.1.1}$$

respectively, for some absolute positive real constants c_{u} and c_{nu} and for all $z \geqslant 0$.

Apparently the best currently known upper bound on c_{u} (in the iid case) is due to Shevtsova [44] and is given by the inequality

$$c_{\mathsf{u}} \leqslant 0.4748. \tag{4.1.2}$$

On the other hand, Esseen's example [8] with iid X_i's, $n \to \infty$, z appropriately close to 0, and

$$\mathsf{P}(X_1 = 1 - p_{\mathsf{Ess}}) = p_{\mathsf{Ess}} = 1 - \mathsf{P}(X_1 = -p_{\mathsf{Ess}}) \tag{4.1.3}$$

with $p_{\mathsf{Ess}} := 2 - \sqrt{10}/2 = 0.4188\ldots$ showed that c_{u} cannot be less than $\frac{3+\sqrt{10}}{6\sqrt{2\pi}} = 0.4097\ldots$. A similar lower bound on the BE constant for

Inequalities and Extremal Problems in Probability and Statistics. http://dx.doi.org/10.1016/B978-0-12-809818-9.00004-5

103

"interval" probabilities of the form $P(S \in I)$ for intervals $I \subseteq \mathbb{R}$ (instead of the probabilities $P(S > Bz))$ was recently shown by Dinev and Mattner [6] to be $\sqrt{\frac{2}{\pi}} = 0.7978\ldots$, which is almost twice as large as $0.4097\ldots$.

Thus, the optimal value of c_u is already known to be within the rather small interval from 0.4097 to 0.4748 in the iid case (in the general, non-iid case the best known upper bound on c_u appears to be 0.5600, due to Shevtsova [46]; a slightly worse upper bound, 0.5606, had been obtained by Tyurin [50]).

4.2 THE BOHMAN–PRAWITZ–VAALER SMOOTHING INEQUALITIES

To a significant extent the mentioned best known uniform BE bounds are based on the smoothing result due to Prawitz [43, (1a, 1b)], which states the following. There exists a nonempty class of functions $M \colon \mathbb{R} \to \mathbb{C}$ such that

$$M(t) = 0 \quad \text{if } |t| > 1 \tag{4.2.1}$$

and for any r.v. X, any real $T > 0$, and any real x,

$$\mathfrak{G}\left(M_T(-\#)\mathsf{E}e^{iX\#}\right)(x) \leqslant P(X < x) - \frac{1}{2} \leqslant P(X \leqslant x) - \frac{1}{2}$$
$$\leqslant \mathfrak{G}\left(M_T(\#)\mathsf{E}e^{iX\#}\right)(x), \tag{4.2.2}$$

where

$$M_T(\#) := M(\#/T), \tag{4.2.3}$$

$$\mathfrak{G}(f)(x) := \frac{i}{2\pi} \,\text{p.v.} \int_{-\infty}^{\infty} e^{-itx} f(t) \frac{dt}{t}, \tag{4.2.4}$$

and p.v. stands for "principal value", so that

$$\text{p.v.} \int_{-\infty}^{\infty} := \lim_{\substack{\varepsilon \downarrow 0 \\ A \uparrow \infty}} \left(\int_{-A}^{-\varepsilon} + \int_{\varepsilon}^{A} \right);$$

here and subsequently, the symbol $\#$ stands for the argument of a function. Of course, the upper and lower bounds in (4.2.2) must take on only real values; this can be provided by the condition that

$$M_1 := \text{Re}\, M \text{ is even} \quad \text{and} \quad M_2 := \text{Im}\, M \text{ is odd.} \tag{4.2.5}$$

Note also that the upper and lower bounds in (4.2.2) easily follow from each other, by changing X to $-X$.

Inequalities (4.2.2) may be compared with the corresponding well-known inversion formula

$$P(X < x) + \frac{1}{2}P(X = x) - \frac{1}{2} = \mathfrak{G}(Ee^{iX\#})(x) \qquad (4.2.6)$$

for all real x; see, for example, [12, (2)].

The function $M(\#)$, whose rescaled version $M_T(\#)$ is the multiplier of the c.f. $Ee^{iX\#}$ in the upper and lower bounds in (4.2.2), is the Fourier transform of the function $\check{M}(\#) := \frac{1}{2\pi} \int_{-\infty}^{\infty} e^{-it\#} M(t)\, dt$, which may be considered as a *bounding smoothing kernel*—since, in view of (4.2.1), the spectral decomposition of $\check{M}$ does not have components of frequencies greater than 1. So, the factors $M(\pm\#/T)$ in the bounds in (4.2.2) filter out the components of the function $\mathfrak{G}(Ee^{iX\#})$ (in (4.2.6)) of frequencies greater than T and thus make the function smoother and flatter, especially if T is not large. Therefore, the inverse Fourier transform $\check{M}(\#)$ of $M(\#)$ may be referred to as a *bounding smoothing filter*.

Another way to look at such smoothing is through the Paley–Wiener theory, which implies that the Fourier spectrum of a function is contained in the interval $[-T, T]$ iff the function is (the restriction to $\mathbb{R}$ of) an entire analytic function of exponential type T and hence rather slowly varying if T is not large; see, for example, [7, Section 43]. On the other hand, from an analytical viewpoint, the presence of the factors $M(\pm\#/T)$ is useful, because one then needs to bound the values Ee^{itX} of the c.f. of X only for $t \in [-T, T]$, which is a much easier task unless T is too large.

One particular bounding smoothing filter M was given by Prawitz [43] and can be defined by the formula

$$M(t) = [(1 - |t|)\, \pi t \cot \pi t + |t| - i(1 - |t|)\, \pi t]\, \mathbb{I}\{|t| < 1\} \qquad (4.2.7)$$

for all $t \neq 0$ [43]; here and subsequently, it is tacitly assumed that the functions of interest are extended to 0 by continuity. For this particular multiplier M, which was shown in [43] to have a certain optimality property, the corresponding smoothing kernel $\check{M}(\#) := \frac{1}{2\pi} \int_{\mathbb{R}} e^{-it\#} M(t)\, dt$ is given by the formula

$$\check{M}(x) = \frac{2\pi x \sin x \left(2\pi(x + 2\pi) - x^2 \psi'\left(\frac{x}{2\pi}\right)\right) - (1 - \cos x)\left(x^3 \psi''\left(\frac{x}{2\pi}\right) + 4\pi^2(x + 4\pi)\right)}{4\pi^3 x^3}.$$

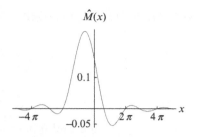

for $x \notin \{-2n\pi : n \in \{0\} \cup \mathbb{N}\}$, where ψ is the digamma function, defined by the formula $\psi(z) = \Gamma'(z)/\Gamma(z)$; this kernel is asymmetric and alternating in sign; also, $\int_{-\infty}^{\infty} \check{M}(x)\,dx = M(0) = 1$; a part of the graph of this kernel $\check{M}$ is shown here on the left.

Remark 4.2.1. The derivative M' of the Prawitz bounding smoothing filter M as in (4.2.7) is a function of bounded variation. So, twice integrating $\int_{\mathbb{R}} e^{-it\#} M(t)\,dt$ by parts and using the Riemann–Lebesgue lemma, one can easily see that the corresponding bounding smoothing kernel $\check{M}$ is such that $x^2 \check{M}(x) - \sin x \to 0$ as $|x| \to \infty$ and hence $\int_{\mathbb{R}} |\check{M}(x)|\,dx < \infty$. Thus, Prawitz's particular M is the Fourier transform of a function $\check{M} \in L^1(\mathbb{R})$.

Earlier, inequalities of the form (4.2.2) were obtained by Bohman [3] for another class of functions M. Another approach to Prawitz's results was demonstrated by Vaaler [52].

4.3 NONUNIFORM BE BOUNDS: NAGAEV'S RESULT AND METHOD

The classical result by Nagaev [21] is that in the "iid" case

$$|P(S > z\sqrt{n}) - P(Z > z)| \leqslant c_{\mathsf{nu}} \frac{\mathsf{E}|X_1|^3}{(1 + z^3)\sqrt{n}} \qquad (4.3.1)$$

for all real $z \geqslant 0$, where c_{nu} is an absolute constant. Bikelis [2] extended this result to the case of non-iid X_i's. Nagaev's method involves the following essential components:

- truncation;
- Cramer's exponential tilt, together with a uniform BE bound;
- an exponential bound on large deviation probabilities.

First, truncated versions of X_i, say $X_i^{(y)}$, are obtained, such that $X_i^{(y)} \leqslant y$ for some real $y > 0$ and all i (the r.v.'s $X_i^{(y)}$ may, in some variants of this approach including [21], be improper in the sense that they may take values that are not real numbers). The truncation is done in order to make the exponential tilt and an exponential inequality possible. The value of the truncation level y is chosen (i) to be large enough so that the tails of the truncated sum $S^{(y)} := X_1^{(y)} + \cdots + X_n^{(y)}$ be close enough to those of S and, on the other hand, (ii) to be small enough so that the exponential tilt and the exponential inequality result in not too large a bound. In some variants, including the ones in [2, 21], two different truncation levels are used.

Take any positive real number z_0. Given a uniform BE bound $c_u\, r_L$ as in (4.1.1), one obviously has the nonuniform BE bound $c_{nu} \frac{r_L}{1+z^3}$ for $z \in (0, z_0)$ and any real $c_{nu} \geqslant c_u(1 + z_0^3)$. So, without loss of generality $z \geqslant z_0$. Two main cases are then considered:

Case 1: $z_0 \leqslant z < c\sqrt{\ln(\sqrt{n}/E|X_1|^3)}$ ("moderate deviations");
Case 2: $z \geqslant z_0 \vee c\sqrt{\ln(\sqrt{n}/E|X_1|^3)}$ ("large deviations");

here c is a positive constant.

In Case 1, of moderate deviations, the exponential tilting is performed, which may be presented as follows. Take some real $h > 0$ and let $\tilde{X}_1 = \tilde{X}_1^{(h,y)}, \ldots, \tilde{X}_n = \tilde{X}_n^{(h,y)}$ be any r.v.'s such that

$$Eg(\tilde{X}_1, \ldots, \tilde{X}_n) = \frac{E\, e^{hS^{(y)}} g(X_1^{(y)}, \ldots, X_n^{(y)})}{E e^{hS^{(y)}}} \qquad (4.3.2)$$

for all bounded (or for all nonnegative) Borel-measurable functions $g\colon \mathbb{R}^n \to \mathbb{R}$. Equivalently, one may require condition (4.3.2) only for Borel-measurable indicator functions g; clearly, such r.v.'s $\tilde{X}_i$ do exist. It is also clear that the r.v.'s $\tilde{X}_i$ are independent. These r.v.'s, the $\tilde{X}_i$'s, may be referred to as the tilted or, more specifically, h-tilted versions of the $X_i^{(y)}$'s. Clearly, without the truncation, the tilted versions of the original r.v.'s X_i may not exist, since Ee^{hS} may be infinite even if $E|X_i|^3 < \infty$ for all i. Using (4.3.2) with $g(x_1, \ldots, x_n) = e^{-h(x_1+\cdots+x_n)} I\{x_1 + \cdots + x_n > x\}$, it is easy to see that

$$P(S^{(y)} > x) = Ee^{hS^{(y)}} \int_x^\infty du\, he^{-hu} P(x < \tilde{S} \leqslant u) \qquad (4.3.3)$$

for all real x, where $\tilde{S} := \tilde{X}_1 + \cdots + \tilde{X}_n$.

Similarly, one can write

$$P(BZ > x) = Ee^{hBZ} \int_x^\infty du\, he^{-hu} P(x < BZ + B^2 h \leqslant u) \qquad (4.3.4)$$

for all real x, since any h-tilted version of the r.v. BZ has the distribution $N(B^2 h, B^2)$.

At that, good choices for y and h are of the form αx and $\eta x/B^2$, for some real parameters α and η in $(0, 1)$.

So, to bound $|P(S^{(y)} > z\sqrt{n}) - P(Z > z)|$ (cf. (4.3.1)), one can demonstrate sufficient closeness of the terms Ee^{hS} and $P(x < \tilde{S} \leqslant u)$ in (4.3.3) to the corresponding terms Ee^{hBZ} and $P(x < BZ + B^2h \leqslant u)$ in (4.3.4).

For each i, one notices that

$$E|X_i^{(y)}|^3 e^{hX_i^{(y)}} \leqslant e^{hy}E|X_i|^3 \qquad (4.3.5)$$

and then shows that $Ee^{hX_i^{(y)}}$ is close enough to 1 and, somewhat more precisely, to $Ee^{hZ\sqrt{EX_i^2}}$, and that the mean and variance of $\tilde{X}_i$ are close enough to hEX_i^2 and EX_i^2, respectively. So, one shows that $Ee^{hS^{(y)}}$ is close to Ee^{hBZ}, and the first two moments of $\tilde{S}^{(y)}$ are close enough to those of $BZ + B^2h$. Using now a uniform BE bound as in (4.1.1)—but for the $\tilde{X}_i$'s rather than the X_i's, one shows that $P(x < \tilde{S} \leqslant u)$ is close enough to $P(x < BZ + B^2h \leqslant u)$.

In Case 2, of large deviations, instead of the exponential tilting and a uniform BE bound, one employs an exponential inequality to bound $P(S^{(y)} > x)$ and hence $P(S^{(y)} > x) - P(BZ > x)$ from above; for the lower bound on the latter difference, one simply uses $-P(BZ > x)$.

4.3.1 A Historical Sketch of the Problem of Nonuniform BE Bounds

The constant factors c_{nu} in the mentioned papers [2, 21] were not explicit. All papers known to this author with explicit values of c_{nu} followed the scheme of proof given by Nagaev [21], as delineated earlier.

Apparently the first such explicit value of c_{nu} was greater than 1955, as reported by Paditz [26]. In his dissertation [25], a much better value, 114.7, was presented. Later, Paditz [27] showed that $c_{nu} < 31.935$.

Michel [17] showed that in the iid case $c_{nu} \leqslant c_u + 8(1+e)$, which would be less than 30.2211, assuming the mentioned value 0.4748 for c_u, obtained in the later paper by Shevtsova [44].

Again in the iid case, Nefedova and Shevtsova [22] briefly stated that they had gone along the lines of the proof in [27] except using a better value for c_u (namely, 0.4784, obtained in [15]) in place of such a value (namely, 0.7915 [47]) used in [27], to get 25.80 for c_{nu}.

Once again in the iid case, Nefedova and Shevtsova claimed in [23] that $c_{nu} < 18.2$. However, there appears to be an error there. Namely, the first inequality in [23, (14)] is equivalent to the reverse of the last inequality on page 75 there, which latter is in turn equivalent to the condition $x^2 \geqslant c_n(x; \delta, a, b, c)$ in [23, Theorem 1], which is also equivalent to the second display on [23, page 75]; the expression $c_n(x; \delta, a, b, c)$ is defined in the first display on page 70 of [23].

So, for any given X, n, b, c, δ satisfying all the conditions of [23, Theorem 1], the first inequality in [23, (14)] and the last inequality on [23, page 75] can both hold only for one value of a. This incorrect inequality in [23, (14)] is also used for [23, (16)].

Working along lines quite similar to those in [23], Grigor'eva and Popov [10, 11] claimed that $c_{nu} < 22.2417$ in the general, non-iid case. However, there appears to be the same kind of errors there: compare [10, (9) and (11)] with [23, (14) and (16)], respectively.

Finally, Shevtsova [45] showed that $c_{nu} \leqslant 17.36$ in the iid case and $c_{nu} \leqslant 21.82$ in the general case.

On the other hand, it follows from a result by Chistyakov [5, Corollary 1] that c_{nu} is necessarily no less than 1, and this lower bound on c_{nu} is asymptotically exact in a certain sense for $z \to \infty$. Apparently, this has been the best known lower bound on c_{nu}. However, it is easy to improve this bound slightly and show that necessarily

$$c_{nu} > 1.0135; \tag{4.3.6}$$

this can be done by letting X_1 have the centered Bernoulli distribution with parameter $p = 8/100$ and then letting $n = 1$ and $z \uparrow 1 - p$.

However, as it was shown by Bentkus [1], the best constant factor for $n = 1$ will be 1 if $1 + z^3$ in (4.3.1) is replaced by z^3; it is also conjectured in [1] that the same constant factor, 1, will be good for all n.

Thus, in the non-iid case the apparently best known lower bound on c_{nu} is over 21 times smaller than the best established upper bound on c_{nu}, and this gap factor is over 17 in the iid case.

4.3.2 Possible Improvements of Nagaev's Method

A crucial component of the mentioned method offered by Nagaev [21] and used in the subsequent papers [2, 9, 17, 18, 22–27, 49] is an exponential inequality. However, the exponential bounds used in all of those papers are not the best possible ones. An optimal exponential bound, in terms of the first two moments and truncated absolute third moments of the X_i's was given by Pinelis and Utev [42]. In fact, the paper [42] provided a general method to obtain optimal exponential bounds, along with a number of specific applications of the general method.

However, even the best possible *exponential* bounds, say for sums of independent r.v.'s, can be significantly improved.

The reason for this is that the class of exponential moments functions is very small (even though analytically very simple to deal with). Using a much richer class of moments functions, Pinelis [31] obtained the following result. Let $X_1, \ldots, X_n$ be independent random variables (r.v.'s), with the sum $S := X_1 + \cdots + X_n$. For any $a > 0$ and $\theta > 0$, let Γ_{a^2} and Π_θ stand for any independent r.v.'s such that Γ_{a^2} has the normal distribution with parameters 0 and a^2, and Π_θ has the Poisson distribution with parameter θ. Let also $\tilde{\Pi}_\theta := \Pi_\theta - \mathsf{E}\Pi_\theta = \Pi_\theta - \theta$. Let σ, y, and β be any positive real numbers such that $\varepsilon := \frac{\beta}{\sigma^2 y} \in (0, 1)$. Suppose that $\sum_i \mathsf{E}X_i^2 \leqslant \sigma^2$, $\sum_i \mathsf{E}(X_i)_+^3 \leqslant \beta$, $\mathsf{E}X_i \leqslant 0$, and $X_i \leqslant y$, for all i. Here and subsequently, we employ the usual notation

$$x_+ := 0 \vee x \quad \text{and} \quad x_- := 0 \wedge x = -(-x)_+.$$

Let $\eta_{\varepsilon,\sigma,y} := \Gamma_{(1-\varepsilon)\sigma^2} + y\tilde{\Pi}_{\varepsilon\sigma^2/y^2}$. Then it is proved in [31] that

$$\mathsf{E}f(S) \leqslant \mathsf{E}f(\eta_{\varepsilon,\sigma,y}) \tag{4.3.7}$$

for all twice continuously differentiable functions f such that f and f'' are nondecreasing and convex. A corollary of this result is that for all $x \in \mathbb{R}$

$$\mathsf{P}\left(S \geqslant x\right) \leqslant \inf_{t \in (-\infty, x)} \frac{\mathsf{E}(\eta_{\varepsilon,\sigma,y} - t)_+^3}{(x - t)^3} \leqslant c_{3,0}\, \mathsf{P}^{\mathsf{LC}}(\eta_{\varepsilon,\sigma,y} \geqslant x), \tag{4.3.8}$$

where $c_{3,0} := \frac{2e^3}{9} \approx 4.46$ and the function $\mathbb{R} \ni x \mapsto \mathsf{P}^{\mathsf{LC}}(\eta \geqslant x)$ is defined as the least log-concave majorant over $\mathbb{R}$ of the tail function $\mathbb{R} \ni x \mapsto \mathsf{P}(\eta \geqslant x)$ of a r.v. η. The bounds in (4.3.7) and (4.3.8) are much better than even the best exponential bounds (expressed in the same terms).

A trade-off here is that the bounds given in (4.3.8) are significantly more difficult to deal with, especially analytically, than exponential bounds. However, this can be done, as shown in the following discussion. In accordance with what was pointed out above, one needs an exponential bound (or a better one) only in Case 2, of large deviations, when $z \geqslant z_0 \vee c\sqrt{\ln(\sqrt{n}/\mathsf{E}|X_1|^3)}$, which implies

$$\mathsf{E}|X_1|^3/\sqrt{n} \geqslant e^{-z^2/c^2}. \tag{4.3.9}$$

Also, by (4.3.8), for any real $\tau < 1$,

$$(1 - \tau)^3 \mathsf{P}(S^{(y)} > x) \leqslant \sum_{j=0}^{\infty} Q_j \frac{\lambda^j}{j!} e^{-\lambda},$$

where

$$x := Bz = z\sqrt{n}, \quad y = \alpha x, \quad \alpha \in (0, 1), \quad Q_j := \left(\frac{\alpha_1}{z} \right)^3 \mathsf{E}(Z + u_j)_+^3,$$

$$\alpha_1 := \sqrt{1 - az_0^2/\alpha}, \quad a := \frac{\mathsf{E}(y \wedge (X_1)_+)^3}{z^3 \sqrt{n}}, \quad u_j := \alpha \left(j - \frac{\tau}{\alpha} - \lambda \right) z,$$

$$\lambda := \frac{a}{\alpha^3}. \tag{4.3.10}$$

Assume now that $\tau c^2 \geqslant 2$, where c is as in (4.3.9). Since $z \geqslant z_0$, one has

$$Q_0 \leqslant C_0 \frac{\mathsf{E}|X_1|^3}{z^3 \sqrt{n}}, \tag{4.3.11}$$

where

$$C_0 := e^{z_0^2/c^2} \mathsf{E}(Z - \tau z_0)_+^3; \tag{4.3.12}$$

here one uses the fact that $e^{\beta t^2} \mathsf{E}(Z - t)_+^3$ is decreasing in $t \geqslant 0$ provided that $\beta \leqslant 1/2$; in fact, this decrease is fast, especially when $\beta < 1/2$. Note also that $\mathsf{E}(Z - t)_+^3 = (t^2 + 2)\,\varphi(t) - t\,(t^2 + 3)\,\bar{\Phi}(t)$ for all real t, where φ and $\bar{\Phi}$ are the density and tail functions of Z.

Next, since $\mathsf{E}g(\beta Z)$ is nondecreasing in $\beta \geqslant 0$ for any convex function g,

$$Q_j \leqslant (\alpha_1/z_0)^3 \mathsf{E}(Z + u_{j0})_+^3, \quad \text{where } u_{j0} := \alpha(j - \tau/\alpha - \lambda)z_0.$$

Using now the identity $\mathsf{E}(Z + t)_+^3 = t^3 + 3t + \mathsf{E}(Z - t)_+^3$ for all real t and the decrease of $\mathsf{E}(Z - t)_+^3$ in $t \in \mathbb{R}$, one has

$$\sum_{1}^{\infty} Q_j \frac{\lambda^j}{j!} e^{-\lambda} \leqslant C_1(a) \frac{\mathsf{E}|X_1|^3}{z^3 \sqrt{n}}, \tag{4.3.13}$$

where

$$C_1(a) := \left(\frac{\alpha_1}{\alpha z_0} \right)^3 \left[\sum_{1}^{\infty} (u_{j0}^3 + 3u_{j0}) \frac{\lambda^{j-1}}{j!} + \mathsf{E}(Z - u_{10})_+^3 \right.$$

$$\left. + \mathsf{E}(Z - u_{20})_+^3 \frac{\lambda^{2-1}}{2!} e^{\lambda} \right] e^{-\lambda}$$

with λ as defined in (4.3.10). The sum $\sum_1^\infty$ in the above expression of $C_1(a)$ is easy to evaluate explicitly. Also, since the left-hand side of (4.3.1) can never exceed 1, without loss of generality

$$a \leqslant a_{max} \qquad (4.3.14)$$

with $a_{max} = 1/c_{nu}$; working a bit harder, one may assume (4.3.14) with a_{max} significantly smaller than $1/c_{nu}$. Next, it appears that for values of the parameters α and τ that have a chance to be optimal or quasioptimal, the factor $C_1(a)$ will be decreasing in $a \in [0, a_{max}]$. Therefore and in view of (4.3.11) and (4.3.13), one will have

$$P(S^{(y)} > x) \leqslant \frac{C_0 + C_1(0+)}{(1 - \tau)^3} \frac{E|X_1|^3}{z^3 \sqrt{n}} \qquad (4.3.15)$$

with C_0 as in (4.3.12) and $C_1(0+) = (\alpha z_0)^{-3} E(Z + (\alpha - \tau)z_0)_+^3$. One can improve the above estimates by partitioning the interval $[0, a_{max}]$ into a number of smaller subintervals and then considering the corresponding cases depending on which of the subintervals the value of a is in.

Thus, it is shown that $P(S^{(y)} > x)$ can be appropriately bounded using the better-than-exponential bound in (4.3.8), and at that in a rather natural manner and incurring almost no losses. It should be clear that the expression on the right-hand side of inequality (4.3.15) will become a term in a bigger expression that is an upper bound on the left-hand side of (4.3.1). That latter, bigger expression will then have to be (quasi)minimized with respect to z_0, α, η, τ, and the other parameters, subject to the necessary restrictions on their values.

Also, one can use ideas from [19, 20, 39, 40] to improve the estimation of the effect of truncation, as compared with the way that was done in the mentioned papers [2, 9, 17, 18, 21–27, 49], as well as more "synthetic" ways to bound moments of the tilted distribution—cf. results in [32, 33, 35, 36].

In addition, as in [23], one can use the uniform bound $0.3328(E|X_1|^3 + 0.429)/\sqrt{n}$ from [44], which is smaller than the previously mentioned bound of the classical form $c_u L = c_u E|X_1|^3/\sqrt{n}$ with $c_u = 0.4748$. There are a few other potentially useful modifications. Thus, the improvements concern every one of the three major ingredients of Nagaev's method listed on page 106. By utilizing the above ideas, one may hope to improve the upper bound on c_{nu} to about 10 in the iid case and to about 12 in the general case. When and if such an objective is attained, the gap between the available upper and lower bounds on c_{nu} will be decreased about 1.5 times.

However, significant further progress after that seems unlikely within the framework of the method of [21]. One of the main obstacles here is the factor e^{hy} as in (4.3.5). Since good choices for y and h turn out to be αx and $\eta x / B^2$ with α and η somewhat close to 0.5, this factor will then be something like $e^{z^2/4}$, which is large for large enough z.

Yet, the factor e^{hy} is the best possible one in (4.3.5) (even assuming that $X_i^{(y)} = X_i$ and hence $X_i^{(y)}$ is zero-mean). Such a large factor is necessary when X_i has a two-point distribution highly skewed to the right. On the other hand, certain considerations suggest that the least favorable situation in Case 1 of moderate deviations is when n is very large but z is not so, and then the mentioned least favorable distribution (for the uniform BE bound) given by (4.1.3) is only slightly skewed. This creates a significant tension in using the exponential tilt.

One may try to reduce the factor e^{hy} by decreasing α and hence y—but this will increase fast the effect of truncation, which is (at least roughly) proportional to $1/\alpha^3$.

Even if one were able to get rid of the factor e^{hy} altogether, the corresponding uniform BE bound on the rate of convergence to the probability $P(x < BZ + B^2 h \leqslant u)$ in (4.3.4) would still seem relatively too large, since this probability itself is less than $P(x < BZ + B^2 h) = P(Z > (1 - \eta)z)$ and therefore is rather small for what appears to be the least favorable values of z, such as 2.5 to 3.5 (and values of η typically not too far from 0.5). In contrast, the mentioned asymptotic lower bound by Esseen [8] (recall (4.1.3)) is attained for z close to 0; furthermore, the corresponding asymptotic expression is rather highly peaked near the maximum and is thus much smaller outside of a neighborhood of the maximum point.

Yet another apparently powerful cause of tension is as follows. After the X_i's have been truncated, a natural bound on $|P(S^{(y)} > x) - P(BZ > x)|$, obtained via either the exponential tilt or a Stein-type method, decays in an exponential rather than power fashion; see, for example, the results [4, 30], which imply an upper bound of the form $c(\lambda) \frac{E|X_1|^3}{e^{\lambda z} \sqrt{n}}$ for real $\lambda > 0$, say in the iid case. The factor $1/e^{\lambda z}$ decays much faster than $1/(1 + z^3)$ when z is large. However, the former factor may be much greater than the latter, especially if λ is not large and z is not very large. For instance, if $\lambda = 1/2$ as in [4], then $\max_{z>0} \frac{1}{e^{\lambda z}} \Big/ \frac{1}{1+z^3} = 10.8\ldots$, attained at $z = 5.9719\ldots$.

In the next section, a new approach to obtaining nonuniform BE bounds is described, based on the Fourier method, complemented by extremal problem methods.

4.4 A NEW WAY TO OBTAIN NONUNIFORM BE BOUNDS

Take any function $h \in C^1$ such that (the limit) $\mathfrak{G}(h)$ exists (and is) in $\mathbb{R}$ and $h(t)/t \to 0$ as $|t| \to \infty$; here and in what follows, C^k denotes the class of all k times continuously differentiable complex-valued functions defined on $\mathbb{R}$. Take any $x \in \mathbb{R}$. Note that $\mathfrak{G}(1)(x) = \frac{1}{2} \operatorname{sign} x$—say, by (4.2.6) with $X = 0$. So, writing $\mathfrak{G}(h) = h(0)\mathfrak{G}(1) + \mathfrak{G}(h - h(0))$ and evaluating the p.v.-integral in the expression for $\mathfrak{G}(h - h(0))$ by parts, one has

$$x\mathfrak{G}(h)(x) = \frac{1}{2} h(0) \, x \operatorname{sign} x + i \, \mathfrak{G}(\Lambda h)(x) \qquad (4.4.1)$$

if $x \neq 0$, where the linear operator Λ is defined by the formula

$$(\Lambda h)(t) := -t \frac{\mathrm{d}}{\mathrm{d}t} \frac{h(t) - h(0)}{t} = -\frac{h(0) - [h(t) + h'(t)(-t)]}{t} \qquad (4.4.2)$$

for $t \neq 0$. In fact, identity (4.4.1) holds for $x = 0$ as well, in view of the definitions of $\mathfrak{G}$ and Λ. By induction, for all $k \in \mathbb{N} := \{1, 2, \dots\}$

$$(\Lambda^k h)(t) = -k! \, t^{-k} \left(h(0) - \sum_{j=0}^{k} h^{(j)}(t) \frac{(-t)^j}{j!} \right)$$

$$= (-1)^k \int_0^1 [h^{(k)}(t) - h^{(k)}(\alpha t)] k\alpha^{k-1} \, \mathrm{d}\alpha \qquad (4.4.3)$$

if $h \in C^k$ and $t \neq 0$, and hence $(\Lambda^k h)(0) = 0$. So, iterating (4.4.1), one has

$$\mathfrak{G}(h)(x) = \frac{h(0) \operatorname{sign} x}{2} + \left(\frac{i}{x} \right)^k \mathfrak{G}(\Lambda^k h)(x) \qquad (4.4.4)$$

for all real $x \neq 0$, all $k \in \mathbb{N}$, and functions h such that

$$h \in C^k, \quad \mathfrak{G}(h) \text{ exists in } \mathbb{R},$$

$$\text{and} \quad h^{(j)}(t)/t \to 0 \text{ for all } j = 1, \dots k \text{ as } |t| \to \infty. \qquad (4.4.5)$$

More generally,

$$\mathfrak{G}(h)(x) = \frac{h(0) \operatorname{sign} x}{2} + \frac{i}{2\pi} \sum_{j=1}^{k} \frac{h^{(j)}(0+) - h^{(j)}(0-)}{j(ix)^j} + \left(\frac{i}{x} \right)^k \mathfrak{G}(\Lambda^k h)(x)$$

$$(4.4.6)$$

for all real $x \neq 0$, all $k \in \mathbb{N}$, and all functions h such that

$h \in C(\mathbb{R})$, $h \in C^k(\mathbb{R} \setminus \{0\})$, $\mathfrak{G}(h)$ exists in $\mathbb{R}$,   and

for each $j \in \{1, \ldots k\}$ there exists $h^{(j)}(0\pm) \in \mathbb{R}$

and $h^{(j)}(t)/t \to 0$ as $|t| \to \infty$.   (4.4.7)

The condition $h \in C^k(\mathbb{R} \setminus \{0\})$ in (4.4.7) can be slightly relaxed, to the following:

$h \in C^{k-1}(\mathbb{R}\setminus\{0\})$ and $h^{(k-1)}$ is of locally bounded variation on $\mathbb{R}\setminus\{0\}$,
                                                                        (4.4.8)

with $\mathfrak{G}(\Lambda^k h)(x)$ then understood as $\mathfrak{G}(\widetilde{\Lambda^k h})(x) + (-1)^k \tilde{\mathfrak{G}}(h^{(k-1)})(x)$, where $(\widetilde{\Lambda^k h})(t) := -k! \, t^{-k} \left(h(0) - \sum_{j=0}^{k-1} h^{(j)}(t) \frac{(-t)^j}{j!} \right)$ for $t \neq 0$ and $\tilde{\mathfrak{G}}(h^{(k-1)})(x) := \frac{i}{2\pi} \, \text{p.v.} \int_{-\infty}^{\infty} e^{-itx} \frac{dh^{(k-1)}(t)}{t}$.

Identity (4.4.4) immediately implies

Theorem 4.4.1. *Take any $k \in \mathbb{N}$ and any real $T > 0$ and $x \geqslant 0$. Let X be any r.v. with $\mathsf{E}|X|^k < \infty$. Let f denote the c.f. of X. Let M be as in (4.2.2), with the additional requirement that $M \in C^k$. Then*

$$- i^k \mathfrak{G}\left(\Lambda^k r_{T,-}\right)(x) \leqslant x^k \mathsf{P}(X > x) \leqslant x^k \mathsf{P}(X \geqslant x) \leqslant -i^k \mathfrak{G}\left(\Lambda^k r_{T,+}\right)(x), \quad (4.4.9)$$

where

$$r_{T,\pm}(\#) := M_T(\mp\#) f(\#). \quad (4.4.10)$$

Remark. Condition $\mathsf{E}|X|^k < \infty$ in Theorem 4.4.1 implies $f \in C^k$, so that (4.4.5) holds with $g = r_{T,\pm}$.

As was mentioned, the Prawitz smoothing filter M given by (4.2.7) provides the tightest, in a certain sense, upper and lower bounds in (4.2.2) on the d.f. of X. However, it is not smooth enough to be used in Theorem 4.4.1 in the most interesting in applications case $k = 3$. Namely, that M is not even in C^1—whereas one needs $M \in C^3$ in Theorem 4.4.1 for $k = 3$. There are a number of ways to develop such a smooth enough smoothing filter. Some of them can be based on Proposition 4.5.1 in Section 4.5 of this paper; see, for example, the function $M = M_{0,2}$ given by formula (4.5.4).

The identity (4.2.6) can be rewritten in the following more general and hence sometimes more convenient form.

Proposition 4.4.2. *Let L be any complex-valued function of bounded variation on $\mathbb{R}$, and let ℓ be its Fourier–Stieltjes transform, so that $\ell(t) = \int_{-\infty}^{\infty} e^{itx}\, dL(x)$ for all real t. Assume also that L is regularized so that $2L(x) = L(x-) + L(x+)$ for all $x \in \mathbb{R}$ and extended to $[-\infty, \infty]$ so that $L(\pm\infty) = \lim_{x \to \pm\infty} L(x)$. Then*

$$L(x) - \frac{1}{2}[L(\infty) - L(-\infty)] = \mathfrak{G}(\ell)(x) \quad \text{for all real } x. \qquad (4.4.11)$$

This follows immediately from (4.2.6), because (i) both sides of (4.4.11) are linear in L and (ii) any regularized function of bounded variation on $\mathbb{R}$ is a linear combination (with complex coefficients) of regularized distribution functions.

Suppose that $\ell \colon \mathbb{R} \to \mathbb{C}$ is a function which may depend on a number of parameters. For brevity, let us say that the function ℓ is a *quasi-c.f.* if it can be represented as a linear combination of k c.f.'s with (possibly complex) coefficients such that the length k of the combination and the coefficients are bounded uniformly over all possible values of the parameters.

Clearly, the product of two quasi-c.f.'s is a quasi-c.f. Also, any linear combination of two quasi-c.f.'s is a quasi-c.f. Moreover, one has the following simple proposition.

Proposition 4.4.3. *Take any natural m. Let N denote the c.f. of a r.v. Y whose distribution may depend on a number of parameters. Suppose that $\mathsf{E}|Y|^m$ is (finite and) bounded uniformly over all possible values of the parameters. Then the mth derivative $N^{(m)}$ of N is a quasi-c.f.*

Proof of Proposition 4.4.3. Let us exclude the trivial case when $\mathsf{E}|Y|^m = 0$. If m is even or $\mathsf{E}Y_+^m = 0$ or $\mathsf{E}Y_-^m = 0$, then $\widetilde{N^{(m)}}(\#) := \frac{\mathsf{E}Y^m e^{iY\#}}{\mathsf{E}Y^m}$ is a c.f., and $N^{(m)} = (i^m \mathsf{E}Y^m)\widetilde{N^{(m)}}$, so that $N^{(m)}$ is a quasi-c.f. In the remaining case one has $\mathsf{E}Y_+^m > 0$ and $\mathsf{E}Y_-^m > 0$, so that one can similarly write $N^{(m)}(\#) = i^m \mathsf{E}Y_+^m \frac{\mathsf{E}Y_+^m e^{iY\#}}{\mathsf{E}Y_+^m} + (-i)^m \mathsf{E}Y_-^m \frac{\mathsf{E}Y_-^m e^{iY\#}}{\mathsf{E}Y_-^m}$. $\qquad\square$

A quick proof of Nagaev's nonuniform BE bound (4.3.1)

It can be easily obtained based on Theorem 4.4.1. Indeed, let $T = c_T \sqrt{n}/\beta_3$, where $\beta_3 := \mathsf{E}|X_1|^3$ and c_T is a small enough positive real constant. Let $A \lesssim B$ mean $|A| \leqslant CB$ for some absolute constant C. Let $X := S/\sqrt{n}$. If $T \leqslant 1$ then $1 \lesssim \frac{\beta_3}{\sqrt{n}}$. So, for all real $x \geqslant 0$, by the Markov and Rosenthal

inequalities, $(1 + x^3)P(X \geqslant x) \leqslant 1 + E|X|^3 \leqslant 1 + \frac{\beta_3}{\sqrt{n}} \leqslant \frac{\beta_3}{\sqrt{n}}$ and
similarly $(1 + x^3)P(Z \geqslant x) \leqslant \frac{\beta_3}{\sqrt{n}}$, whence (4.3.1) follows. It remains to
consider the case $T > 1$. Note that then $n > (\beta_3/c_T)^2 \geqslant 3$ and hence
$n \geqslant 4$ provided that $c_T \leqslant 1/\sqrt{3}$. In view of the uniform BE bound,
Theorem 4.4.1, and (4.4.3), in order to prove (4.3.1) it is enough to show
that $\mathfrak{G}\left(r_{1,f}'''(\alpha\#) - r_{1,g}'''(\alpha\#) \right) \leqslant \beta_3/\sqrt{n}$ and $\mathfrak{G}\left(r_{2,f}'''(\alpha\#) \right) \leqslant \beta_3/\sqrt{n}$
over $\alpha \in (0, 1]$, where $r_{j,f}(\#) := M_j(\frac{\#}{T})f(\#)$, $j = 1, 2$, the M_j's are
as in (4.2.5), M is (say) as in (4.5.4), f is the c.f. of $X := S/\sqrt{n}$, and
$g(\#) := e^{-\#^2/2}$ (so that g may be considered as a special case of f).
One has

$$r_{j,f}'''(\#) = \sum_{q=0}^{3} \binom{3}{q} \frac{1}{T^q} M_j^{(q)}\left(\frac{\#}{T} \right) f^{(3-q)}(\#). \tag{4.4.12}$$

By (4.5.4) and (4.5.6), M_1 is the c.f. of a distribution with a finite fourth moment, whereas $M_2 = \kappa M_1'$. Hence, by Proposition 4.4.3, $M_j^{(q)}$ is a quasi-c.f.
for each pair $(j, q) \in \{1, 2\} \times \{0, 1, 2, 3\}$, and then so is $M_j^{(q)}(\frac{\#}{T})$. Similarly,
f is the c.f. of the r.v. X with $E|X|^3 \leqslant 1 + \frac{\beta_3}{\sqrt{n}} \leqslant 1$, by the Rosenthal
inequality and the case condition $T > 1$. So, again by Proposition 4.4.3,
$f^{(3-q)}$ is a quasi-c.f. for each $q \in \{0, 1, 2, 3\}$. Thus, $M_j^{(q)}(\frac{\alpha\#}{T})f^{(3-q)}(\alpha\#)$
is a quasi-c.f. and, by Proposition 4.4.2, $\mathfrak{G}\left(M_j^{(q)}\left(\frac{\alpha\#}{T} \right) f^{(3-q)}(\alpha\#) \right) \leqslant 1$,
for each $(j, q) \in \{1, 2\} \times \{0, 1, 2, 3\}$. Therefore and because $T > 1$,

$$\mathfrak{G}\left(\binom{3}{q} \frac{1}{T^q} M_j^{(q)}\left(\frac{\alpha\#}{T} \right) f^{(3-q)}(\alpha\#) \right) \leqslant \frac{1}{T^q} \leqslant \frac{1}{T}$$

$$\leqslant \frac{\beta_3}{\sqrt{n}} \quad \text{for each } (j, q) \in \{1, 2\} \times \{1, 2, 3\}; \tag{4.4.13}$$

note that $q = 0$ is not included here.

It remains to show that $\mathfrak{G}_{1\alpha}(f''' - g''') \leqslant \frac{\beta_3}{\sqrt{n}}$ and $\mathfrak{G}_{2\alpha}(f''') \leqslant \frac{\beta_3}{\sqrt{n}}$ for
$\alpha \in (0, 1]$, where

$$\mathfrak{G}_{j\alpha}(h)(x) := \mathfrak{G}\left(M_j\left(\frac{\alpha\#}{T} \right) h(\alpha\#) \right)(x). \tag{4.4.14}$$

For $j \in \{0, 1, 2, 3\}$, introduce $f_1^{(j)}(t) := \left(\frac{d}{dt} \right)^j f_1(t)$ and $f_{1n}^{(j)}(t) :=$
$f_1^{(j)}(t/\sqrt{n})$, where f_1 denotes the c.f. of X_1.

Similarly, starting with $g_1 := g$ in place of f_1, define $g_{1n}^{(j)}$, and then let $d_{1n}^{(j)} := f_{1n}^{(j)} - g_{1n}^{(j)}$ and $h_{1n}^{[j]} := \left| f_{1n}^{(j)} \right| \vee \left| g_{1n}^{(j)} \right|$; omit superscripts (0) and $[0]$. Note that $f = f_{1n}^n$ and hence $\sqrt{n} f''' = f_{31} + f_{32} + f_{33}$, where

$$f_{31} := (n-1)(n-2) f_{1n}^{n-3} \left(f_{1n}^{(1)} \right)^3, \quad f_{32} := 3(n-1) f_{1n}^{n-2} f_{1n}^{(1)} f_{1n}^{(2)},$$

and $f_{33} := f_{1n}^{n-1} f_{1n}^{(3)}$; do similarly with g and g_1 in place of f and f_1. By Proposition 4.4.3, $M_j \left(\frac{\alpha\#}{T} \right) f_{33}/\beta_3$ is a quasi-c.f. and hence, by Proposition 4.4.2, $\mathfrak{G}_{j\alpha}(f_{33}) \lesssim \beta_3$, for $j \in \{1,2\}$.

So, it suffices to show that $\mathfrak{G}_{1\alpha}(f_{3k} - g_{3k}) \lesssim \beta_3$ and $\mathfrak{G}_{2\alpha}(f_{3k}) \lesssim \beta_3$ for $k \in \{1,2\}$. This can be done in a straightforward manner using the following estimates for $j \in \{0,1,2,3\}$ and $|t| \leq T$: $M_1 \leq 1$, $M_2(\frac{t}{T}) \leq \frac{|t|}{T} \leq |t|\beta_3/\sqrt{n}$, $h_{1n}(t)^{n-j} \leq e^{-ct^2}$ (where c is a positive real number depending only on the choice of c_T), $h_{1n}^{[1]}(t) \leq |t|/\sqrt{n}$, $h_{1n}^{[2]}(t) \leq 1$, $|d_{1n}^{(j)}(t)| \leq \beta_3(|t|/\sqrt{n})^{3-j}$, and hence $f_{1n}^{n-j}(t) - g_{1n}^{n-j}(t) \leq |t|^3 e^{-ct^2} \beta_3/\sqrt{n}$; cf., for example, [28, Ch. V, Lemma 1]. For instance, $|f_{31} - g_{31}| \leq n^2(D_{311} + D_{312})$, where $D_{311}(t) := \left(|f_{1n}^{n-3} - g_{1n}^{n-3}| (h_{1n}^{[1]})^3 \right)(t) \leq |t|^3 e^{-ct^2} \frac{\beta_3}{\sqrt{n}} \left(\frac{|t|}{\sqrt{n}} \right)^3$ and $D_{312}(t) := \left(h_{1n}^{n-3} (h_{1n}^{[1]})^2 |d_{1n}^{(1)}| \right)(t) \leq e^{-ct^2} \left(\frac{|t|}{\sqrt{n}} \right)^2 \beta_3 \left(\frac{|t|}{\sqrt{n}} \right)^2$, so that $\mathfrak{G}_{1\alpha}(f_{31} - g_{31}) \lesssim \int_{-\infty}^{\infty} (t^6 + t^4) e^{-ct^2} \beta_3 \frac{dt}{|t|} \lesssim \beta_3$.

Of course, the above argument is rather crude and yet it demonstrates that the method based on the smoothing inequalities (4.4.9) is quite effective. It also strongly suggests that this method can be used further, in order to obtain an explicit and appropriately small upper bound on the constant factor c_{nu}.

Let us now discuss some of the refinements that could be used within the general framework of the above *quick proof of* (4.3.1). There, in particular, we needed to bound

$$L(H) := \int_0^1 [\mathfrak{G}(H(\alpha\#))(x) - \mathfrak{G}(H(\#))(x)] 3\alpha^2 \, d\alpha, \qquad (4.4.15)$$

where H is of the form $M_1(\frac{\#}{T})(f_{3k} - g_{3k})$ or $M_2(\frac{\#}{T}) f_{3k}$ for $k \in \{1,2\}$— recall (4.4.9), (4.4.10), and (4.4.3). Tacitly, that bounding was then done using the trivial inequalities

$$|L(H)| \leq 2 \sup_{\alpha \in (0,1]} |\mathfrak{G}(H(\alpha\#))(x)| \leq 2 \frac{1}{2\pi} \int_{-\infty}^{\infty} |H(t)| \frac{dt}{|t|}, \qquad (4.4.16)$$

where in turn we used the definition (4.2.4) of $\mathfrak{G}$ and the trivial identity $|e^{-itx}| = 1$ for real t and x; the integral in (4.4.16) exists even in the Lebesgue sense, since $|f_{3k}(t) - g_{3k}(t)| = O(|t|)$ and $|M_2(t)| = O(|t|)$.

In fact, the factor 2 in the last bound in (4.4.16) on $|L(H)|$ can be removed, so that one has

$$|L(H)| \leqslant \frac{1}{2\pi} \int_{-\infty}^{\infty} |H(t)| \frac{dt}{|t|}. \tag{4.4.17}$$

Indeed, first of all note here that the factor α can be easily moved, in a way, from the argument of the general and hard to control function H into that of the much simpler and more specific exponential function, using the simple identity

$$\mathfrak{G}\left(H(\alpha\#)\right)(x) = \mathfrak{G}\left(H(\#)\right)\left(\frac{x}{\alpha}\right), \tag{4.4.18}$$

which implies that

$$L(H) = \frac{i}{2\pi} \int_{-\infty}^{\infty} I(tx)H(t)\frac{dt}{t}, \tag{4.4.19}$$

where

$$I(u) := \int_0^1 \left(e^{-iu/\alpha} - e^{-iu}\right) 3\alpha^2 \, d\alpha. \tag{4.4.20}$$

Now (4.4.17) follows immediately from

Proposition 4.4.4. *The expression $g(u) := |I(u)|^2$ is even in $u \in \mathbb{R}$ and (strictly) increases from 0 to 1 as $|u|$ increases from 0 to ∞; in particular, it follows that $|I(u)| \in [0, 1)$ for all real u. Moreover, the function g has the following generalized concavity property: $-u^3 \left(u^{-5}g'(u)\right)'$ is completely monotone in $u > 0$ (in Bernstein's sense—see, e.g., [29, Chapter 2]); in particular, $g(v^{1/6})$ is concave in $v > 0$.*

Thus, the conclusion in Proposition 4.4.4 that $|I(u)| \in [0, 1)$ for all real u can be seen as a rather sophisticated replacement for the trivial identity $|e^{-itx}| = 1$ for real t and x, which latter was used to obtain the rightmost bound in (4.4.16).

Moreover, one can easily obtain (and then use in (4.4.19)) an upper bound on $|I(u)|$ which is significantly less than 1 for small enough values of $|u|$. This can be done by closely bounding the values of $|I(u)|$ for a finite number of values of u and then using the monotonicity property of $|I|$ provided by Proposition 4.4.4.

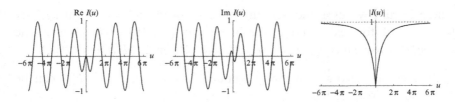

Fig. 4.1 Graphs of Re I, Im I, *and* $|I|$.

Graphs of Re I, Im I, and $|I|$ over the interval $[-6\pi, 6\pi]$ are shown in Fig. 4.1. It seems plausible that $g(u)$ is concave in $u > 0$; however, that probably would be hard to prove.

Proof of Proposition 4.4.4. Note that $I(-u) = \overline{I(u)}$ for all real u. So, the function $g = I\overline{I}$ is indeed even.

Take now any real $u > 0$. Integrating by parts and then changing the integration variable, one has

$$iu^{-3}I(u)$$

$$= u^{-2} \int_0^1 e^{-iu/\alpha} \, \alpha \, d\alpha = \mathcal{E}(u) := \mathcal{E}_3(u), \quad \text{where } \mathcal{E}_j(u) := \int_u^\infty e^{-iz} \frac{dz}{z^j} \tag{4.4.21}$$

for $j > 0$. So, $g(u) = u^6 \, \mathcal{E}(u)\overline{\mathcal{E}(u)}$,

$$g_1(u) := \frac{1}{2} u^{-5} g'(u) = 3\mathcal{E}(u)\overline{\mathcal{E}(u)} - u^{-2} \operatorname{Re}\left(e^{iu}\mathcal{E}(u)\right),$$

$$u^3 g_1'(u) = \frac{1}{u^2} - \operatorname{Re}\left((4+iu)e^{iu}\mathcal{E}(u)\right)$$

$$= \frac{1}{u^2} - \operatorname{Re}\left((4+iu)\int_u^\infty e^{-i(z-u)} \frac{dz}{z^3}\right)$$

$$= \frac{1}{u^2} - \operatorname{Re}\left((4+iu)\int_0^\infty e^{-iv} \frac{dv}{(v+u)^3}\right)$$

$$= \int_0^\infty e^{-us} s \, ds - \operatorname{Re}\left((4+iu)\int_0^\infty e^{-iv} \, dv \int_0^\infty \frac{1}{2} e^{-(v+u)s} s^2 \, ds\right).$$

Noting now that $\int_0^\infty e^{-iv}e^{-vs} \, dv = \frac{1}{s+i}$ and $\operatorname{Re} \frac{4+iu}{s+i} = \frac{4s+u}{s^2+1}$ for all real $s > 0$, and introducing $w_1(s) := 2\frac{s-s^3}{s^2+1}$ and $w_2(s) := \frac{-s^2}{s^2+1}$, write

$$2u^3 g_1'(u) = \int_0^\infty e^{-us} w_1(s)\, ds + \int_0^\infty e^{-us} w_2(s)\, u\, ds$$

$$= \int_0^\infty e^{-us} w_1(s)\, ds + \int_0^\infty e^{-us} w_2'(s)\, ds = -2 \int_0^\infty e^{-us} \frac{s^5\, ds}{(s^2+1)^2},$$

which verifies the last sentence of the statement of Proposition 4.4.4; the second equality in the above display was obtained by taking the integral $\int_0^\infty e^{-us} w_2(s)\, u\, ds$ by parts. Moreover, it follows that $g_1(u)$ is decreasing in $u > 0$. At that, $g_1(\infty-) = 0$, since $\mathcal{E}(\infty-) = 0$. So, on $(0, \infty)$ one has the following: $g_1 > 0$ and hence $g' > 0$, and therefore g is increasing. Since g is even and obviously continuous, it follows that indeed $g(u)$ increases in $|u|$. Clearly, $g(0) = |I(0)|^2 = 0$. It remains only to show that $g(\infty-) = 1$. Toward that end, integrate by parts to obtain the recursive relation $\mathcal{E}_j(u) = -ie^{-iu}u^{-j} + ij\mathcal{E}_{j+1}(u)$ for all $j > 0$. In particular, it follows that $\mathcal{E}_3(u) = -ie^{-iu}u^{-3} + 3i\mathcal{E}_4(u)$ and $\mathcal{E}_4(u) \leqslant u^{-4} + |\mathcal{E}_5(u)| \leqslant u^{-4} + \int_u^\infty \frac{dz}{z^5} \leqslant u^{-4}$. Thus, $\mathcal{E}(u) = \mathcal{E}_3(u) = -ie^{-iu}u^{-3} + O(u^{-4}) = -iu^{-3}\left(e^{-iu} + o(1)\right)$ and $g(u) = u^6 |\mathcal{E}(u)|^2 \to 1$ as $u \to \infty$. $\qquad\square$

One could also use a better upper bound on $|f(t)|$ for a given real value of t, where f is the c.f. of a r.v. X, say with $\mathsf{E}X = 0$, $\mathsf{E}X^2 = 1$, and a given value of $\rho := \mathsf{E}|X|^3$. Since

$$|f(t)| = \sqrt{\mathsf{E}^2 \cos tX + \mathsf{E}^2 \sin tX} = \sup_{\theta \in [0,2\pi]} (\cos\theta\, \mathsf{E}\cos tX + \sin\theta\, \mathsf{E}\sin tX)$$

$$= \sup_{\theta \in [0,2\pi]} \mathsf{E}\cos(tX - \theta), \tag{4.4.22}$$

the best upper bound on $|f(t)|$ under the given conditions is

$$S(t, \rho) := \sup\{|f(t)| \colon \mathsf{E}X = 0, \mathsf{E}X^2 = 1, \mathsf{E}|X|^3 = \rho\} \tag{4.4.23}$$

$$= \sup\{\mathsf{E}\cos(tX - \theta) \colon \mathsf{E}X = 0, \mathsf{E}X^2 = 1, \mathsf{E}|X|^3 = \rho,$$

$$\operatorname{card\,supp} X \leqslant 4, \theta \in [0, 2\pi]\}$$

$$= \sup\left\{ \sum_1^4 p_j \cos(tx_j - \theta) \colon \sum_1^4 p_j = 1, \sum_1^4 p_j x_j = 0, \right.$$

$$\sum_1^4 p_j x_j^2 = 1, \sum_1^4 p_j |x_j|^3 = \rho,$$

$$\left. p_1, \ldots, p_4 \geqslant 0,\ \theta \in [0, 2\pi] \right\};$$

card supp denotes the cardinality of the support of (the distribution of) X; for the second equality here, one can use the known results by Hoeffding [13] or Karr [14] or, somewhat more conveniently, Winkler [53] or Pinelis [38, Propositions 5 and 6(v)]. Thus the optimization problem reduces to one in nine variables: $p_1, \ldots, p_4, x_1, \ldots, x_4, \theta$; in fact, one can easily solve the linear (or, more precisely, affine) restrictions $\sum_1^4 p_j = 1, \sum_1^4 p_j x_j = 0$, $\sum_1^4 p_j x_j^2 = 1, \sum_1^4 p_j |x_j|^3 = \rho$ for $p_1, \ldots, p_4$, and then only five variables will remain: $x_1, \ldots, x_4, \theta$, with the additional restrictions on $x_1, \ldots, x_4$ to provide for the conditions $p_1, \ldots, p_4 \geqslant 0$. For any given pair of values of (t, ρ), it will not be overly hard to find a close upper bound on the supremum $\mathcal{S}(t, \rho)$. A difficulty here is that one has to deal with two parameters, t and ρ, and obtain a close majorant of $\mathcal{S}(t, \rho)$ with, at least, discoverable and tractable patterns of monotonicity/convexity in t and ρ, if not with a more or less explicit expression. Apparently the main difficulty in dealing with $\mathcal{S}(t, \rho)$ will be that the target function $\cos(t\# - \theta)$ oscillates, whereas the function $(\# - w)_+^3$ in (say) [37, Lemma 3.4] is monotonic.

Similar methods can be used to find a good upper bound on $|f(t) - g(t)|$, where $g = e^{-\#^2/2}$, the c.f. of the standard normal r.v. Z; in particular, one can start here by writing

$$|f(t) - g(t)| = \sup_{\theta \in [0, 2\pi]} \left(\mathsf{E} \cos(tX - \theta) - \mathsf{E} \cos(tZ - \theta) \right)$$
$$= \sup_{\theta \in [0, 2\pi]} \left(\mathsf{E} \cos(tX - \theta) - g(t) \cos \theta \right)$$

in place of (4.4.22). At this point, one also has an option to use Stein's method to bound $\mathsf{E} \cos(tX - \theta) - \mathsf{E} \cos(tZ - \theta)$.

4.5 CONSTRUCTIONS OF THE SMOOTHING FILTER M

The following proposition was somewhat implicit in the paper [3] by Bohman.

Proposition 4.5.1. *Let p be any symmetric probability density function (p.d.f.) such that the function $\#p(\#)$ is integrable on $\mathbb{R}$. Take any real*

$$\kappa \geqslant \kappa_* := \frac{1}{\int_{\mathbb{R}} |x| p(x) \, dx}. \tag{4.5.1}$$

Let $\hat{p}$ stand, as usual, for the Fourier transform of p (so that $\hat{p}(\#) = \int_{\mathbb{R}} e^{ix\#} p(x) \, dx$), and let then $\hat{p}'$ denote the derivative of $\hat{p}$ (which exists, since $\int_{\mathbb{R}} |x| p(x) \, dx < \infty$). Then the function

$$M := \hat{p} + i\kappa \hat{p}' \tag{4.5.2}$$

is such that inequalities (4.2.2) hold for all r.v. X, all real $T > 0$, and all real x.

Because of the symmetry of p, in the conditions of Proposition 4.5.1 the function $\hat{p} = \text{Re}\, M$ is even and $\kappa\hat{p}' = \text{Im}\, M$ is odd, so that conditions (4.2.5) hold. In order to satisfy the conditions (4.2.1) and $M \in C^3$ as well, one may choose the symmetric p.d.f. $p_{0,2}$ defined by the formula

$$p_{0,2}(x) := \frac{32\pi^3}{3}\,\frac{1 - \cos x}{x^2\,(x^2 - 4\pi^2)^2} \tag{4.5.3}$$

for real $x \notin \{-2\pi, 0, 2\pi\}$ and then let M be as in (4.5.2) with $p = p_{0,2}$:

$$M = M_{0,2} := \widehat{p_{0,2}} + i\kappa\widehat{p_{0,2}}' \tag{4.5.4}$$

with any

$$\kappa \geqslant \kappa_{0,2} := \frac{1}{\int_\mathbb{R} |x|p_{0,2}(x)\,\mathrm{d}x} = 0.3418\ldots. \tag{4.5.5}$$

Then $M_{0,2} \in C^3$, since $M''' = \hat{p}''' + i\kappa\hat{p}''''$ and

$$\int_\mathbb{R} x^4\, p_{0,2}(x)\,\mathrm{d}x < \infty. \tag{4.5.6}$$

Moreover, it is clear that $p_{0,2}$ is the restriction to $\mathbb{R}$ of an entire analytic function of exponential type 1; so, by the Paley–Wiener theory (see, e.g., [7, Section 43]), the condition (4.2.1) holds as well, with $M_{0,2}$ in place of M. In fact,

$$\text{Re}\, M_{0,2}(t) = \widehat{p_{0,2}}(t) = \left(\frac{2 + \cos 2\pi t}{3}\,(1 - |t|) + \frac{\sin 2\pi|t|}{2\pi}\right) \text{I}\{|t| < 1\} \text{ and}$$

$$\tag{4.5.7}$$

$$\text{Im}\, M_{0,2}(t) = \kappa\widehat{p_{0,2}}'(t)$$

$$= -\kappa\,\frac{\text{sign}\, t}{3}\,\left[2\pi(1 - |t|)\sin 2\pi|t| + 4\sin^2(\pi t)\right] \text{I}\{|t| < 1\}$$

$$\tag{4.5.8}$$

for all real t. Graphs of $p_{0,2}$, $\text{Re}\, M_{0,2}$, and $\text{Im}\, M_{0,2}$ with $\kappa = \kappa_{0,2}$ are shown in Fig. 4.2.

More generally, in order that a function M as in (4.5.2) satisfy the conditions (4.2.1) and $M \in C^3$, it suffices that $\hat{p}$ be smooth enough and such that (4.2.1) holds with $\hat{p}$ in place of M. Therefore, the following well-known characterization is useful.

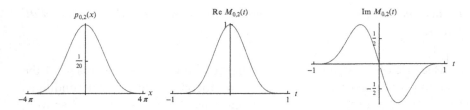

Fig. 4.2 Graphs of $p_{0,2}$, Re $M_{0,2}$, and Im $M_{0,2}$ with $\kappa = \kappa_{0,2}$.

Proposition 4.5.2. *[See, e.g., [16, Theorem 4.2.4]] A function $f \colon \mathbb{R} \to \mathbb{C}$ is the c.f. of an absolutely continuous distribution on $\mathbb{R}$ if and only if $f(0) = 1$ and $f = g * \overline{g}^-$ for some (possibly complex-valued) function $g \in L^2(\mathbb{R})$. Here and in the sequel, as usual, the symbol $*$ stands for the convolution, the bar denotes the complex conjugation, $g^-(\#) := g(-\#)$, and $\overline{g}^- := (\overline{g})^- = \overline{g^-}$.*

Indeed, take any smooth enough nonzero function $g \colon \mathbb{R} \to \mathbb{C}$ such that $g(t) = 0$ for all real $t \notin [a, b]$, where a and b are any real numbers such that $a < b$. Then, by Proposition 4.5.2, the function $f := g * \overline{g}^- / \|g\|_2^2$ is the c.f. of an absolutely continuous distribution on $\mathbb{R}$, f is smooth enough, and $f(t) = 0$ for all real $t \notin [-T, T]$, where $T := b - a$. At that, if g is real-valued, then f is even. To spell-out the "smooth enough" condition and conclusion here, one can easily check that, if $g \in C^j$ and $h \in C^k$ for some j and k in $\mathbb{N}$ and (say) $|g(t)| + |h(t)| = 0$ for some real $T > 0$ and all real $t \notin [0, T]$, then $g * h \in C^{j+k}$, with $(g * h)^{(j+k)} = g^{(j)} * h^{(k)}$.

One can use Propositions 4.5.1 and 4.5.2 to optimize properties of the filter M—say by taking g to be an arbitrary nonzero real-valued spline of a high enough order and/or with a large enough subintervals of the interval $[0, T]$, extending g to $\mathbb{R}$ by letting $g(t) := 0$ for all real $t \notin [0, T]$, letting then $f := g * \overline{g}^- / \|g\|_2^2$, defining M as in (4.5.2), and finally (quasi)optimizing with respect to the parameters of the spline.

While the construction described in Proposition 4.5.1 is comparatively simple, it appears somewhat too rigid and wasteful. Indeed, in order that the imaginary part $M_2 = \kappa \hat{p}'$ of the function M in (4.5.2) be thrice differentiable (as needed or almost needed in the *quick proof* beginning on page 116), the real part $M_1 = \hat{p}$ of M must be four times differentiable; equivalently (cf. (4.5.6)), the density p must have light enough tails so that $\int_{\mathbb{R}} x^4 p(x) \, \mathrm{d}x < \infty$. Together with the filtering condition (4.2.1), the condition of extra smoothness of $M_1 = \hat{p}$/extra lightness of the tails of p may result in a smoothing filter M which is not as good as it can be, thus compromising the quality of the approximation by the upper and lower bounds in (4.4.9).

A more flexible and potentially better construction of the smoothing filter M can be given as follows. As in Proposition 4.5.1, let us start with an arbitrary symmetric p.d.f. p, whose Fourier transform $\hat{p}$ is intended to be $M_1 = \operatorname{Re} M$. Accordingly, let us assume right away that

$$\hat{p}(t) = 0 \quad \text{if } |t| > 1; \tag{4.5.9}$$

cf. (4.2.1). Note that the smoothing filter as in (4.5.2) is the Fourier transform of the function

$$x \mapsto p(x)(1 - \kappa x), \tag{4.5.10}$$

which differs relatively much from the "original" p.d.f. $x \mapsto p(x)$ when $|x|$ is large.

To address this concern, let us replace the "large" factor x in (4.5.10) by a "tempered" and, essentially, more general factor $G(x)$ such that $G \colon \mathbb{R} \to \mathbb{R}$ is a strictly increasing odd function of bounded variation, whose Fourier–Stieltjes transform $\widehat{dG}(\#) = \int_{\mathbb{R}} e^{ix\#}\, dG(x)$ satisfies the condition

$$\widehat{dG}(t) = 0 \quad \text{if } |t| > \gamma, \tag{4.5.11}$$

for some real $\gamma > 0$. The no-high-frequency-component condition (4.5.11) implies, by the mentioned Paley–Wiener theory, that the function G is the restriction to $\mathbb{R}$ of an entire analytic function of exponential type γ and, in particular, is infinitely many times differentiable. Without loss of generality, assume that the function $\frac{1}{2} + G$ is a d.f.

As mentioned earlier, instead of the "harsh" tilting (4.5.10) of the p.d.f. p, we consider the "tempered" tilting:

$$x \mapsto \tilde{p}(x) := p(x)\,(1 - \kappa G(x)), \tag{4.5.12}$$

for any real

$$\kappa \geqslant \kappa_* := \frac{1}{2 \int_0^\infty p(x) G(x)\, dx}. \tag{4.5.13}$$

Note that $0 \leqslant G(x) < \frac{1}{2} = G(\infty-)$ for all real $x > 0$; also, as discussed previously, the condition (4.5.9) implies that p is the restriction to $\mathbb{R}$ of an entire analytic function, and so, $p > 0$ almost everywhere on $\mathbb{R}$. Therefore and by the symmetry of p, one has $0 < \int_0^\infty p(x)G(x)\,dx < \frac{1}{2}\int_0^\infty p(x)\,dx = \frac{1}{4}$ and hence $\kappa_* > 2$ and $\kappa > 2$. It follows that there exists a unique root $x_\kappa \in (0, \infty)$ of the equation

$$1 - \kappa G(x_\kappa) = 0. \qquad (4.5.14)$$

Hence, $\tilde{p} \geqslant 0$ on the interval $(-\infty, x_\kappa]$ and $\tilde{p} \leqslant 0$ on $[x_\kappa, \infty)$, so that the function

$$\tilde{F}(\#) := \int_{-\infty}^{\#} \tilde{p}(y)\,dy \qquad (4.5.15)$$

is nondecreasing on $(-\infty, x_\kappa]$ and nonincreasing on $[x_\kappa, \infty)$. At that, $\tilde{F}(\infty-) = 1$, since p is an even p.d.f. and the bounded function G is odd; also, clearly $\tilde{F}(-\infty+) = 0$. Moreover, $\tilde{F}(0) = \frac{1}{2} - \kappa \int_{-\infty}^0 p(x)G(x)\,dx = \frac{1}{2} + \kappa \int_0^\infty p(x)G(x)\,dx = \frac{1}{2} + \frac{\kappa}{2\kappa_*} \geqslant 1$. It follows that

$$\mathrm{I}\{y \geqslant 0\} \leqslant \tilde{F}(y) \quad \text{for all real } y. \qquad (4.5.16)$$

Let now X be any r.v. and let f be its c.f.:

$$f(\#) := \mathsf{E}e^{iX\#}. \qquad (4.5.17)$$

Then $\mathsf{P}(X \leqslant x) = \mathsf{E}\,\mathrm{I}\{x - X \geqslant 0\} \leqslant \mathsf{E}\tilde{F}(x - X)$, by (4.5.16); that is,

$$\mathsf{P}(X \leqslant x) \leqslant \int_{\mathbb{R}} \tilde{F}(x - y)\mathsf{P}(X \in dy), \qquad (4.5.18)$$

for all real x. Define now M as the Fourier transform of $\tilde{p}$, so that

$$M = \hat{\tilde{p}}, \quad M_1 = \operatorname{Re} M = \hat{p}, \quad \text{and} \quad M_2 = \operatorname{Im} M = i\kappa \widehat{pG}, \qquad (4.5.19)$$

by (4.5.12). Note that the Fourier–Stieltjes transform of the function $\int_{\mathbb{R}} \tilde{F}(\# - y)\mathsf{P}(X \in dy)$ is the Fourier transform of $\int_{\mathbb{R}} \tilde{p}(\# - y)\mathsf{P}(X \in dy)$, which in turn is $\hat{\tilde{p}}f = Mf$. Then, in view of Proposition 4.4.2, (4.5.18) means that the last inequality in (4.2.2) holds for $T = 1$; that it holds for any real $T > 0$ now follows by simple rescaling, since (4.5.16) obviously implies $\mathrm{I}\{y \geqslant 0\} \leqslant \tilde{F}(Ty)$ for all real y and all $T > 0$. Similarly or using the reflection $x \mapsto -x$, one can see that the first inequality in (4.2.2) holds as well.

To compute M_2 in (4.5.19), we need to express $\widehat{pG}$ in terms of $\hat{p}$ and $\widehat{dG}$. To simplify the derivation, assume the condition $\hat{p} \in C^1$, as well as the previously stated conditions (4.5.9) and (4.5.11); these conditions will hold in the applications anyway. Then one can see that for all real u

$$\widehat{pG}(u) = \frac{i}{2\pi} \,\text{p.v.} \int_{-\infty}^{\infty} \hat{p}(u-s)\,\widehat{dG}(s)\frac{ds}{s}$$

$$= \frac{i}{2\pi} \int_{\mathbb{R}} \frac{\hat{p}(u-s) - \hat{p}(u)}{s}\,\widehat{dG}(s)\,ds; \tag{4.5.20}$$

the latter equality here holds because the function G was assumed odd, and hence the function $\widehat{dG}$ is even; the latter integral in (4.5.20) may be understood in the Lebesgue sense, in view of the conditions $\hat{p} \in C^1$ and (4.5.11). It follows from (4.5.20), (4.5.9), and (4.5.11) that

$$\widehat{pG}(t) = 0 \quad \text{if } |t| > 1 + \gamma. \tag{4.5.21}$$

To verify the first equality in (4.5.20), one can write

$$p(x)G(x) = \left(\frac{1}{2\pi}\int_{\mathbb{R}} e^{-itx}\hat{p}(t)\,dt\right)\left(\frac{i}{2\pi}\,\text{p.v.}\int_{-\infty}^{\infty} e^{-isx}\,\widehat{dG}(s)\frac{ds}{s}\right)$$

$$= \frac{1}{2\pi}\int_{\mathbb{R}} e^{-iux}\,du\,\frac{i}{2\pi}\,\text{p.v.}\int_{-\infty}^{\infty} \hat{p}(u-s)\,\widehat{dG}(s)\frac{ds}{s};$$

the second equality here is justified because of the second equality in (4.5.20) and the inequality $\left|\int_{\mathbb{R}} \frac{\hat{p}(u-s)-\hat{p}(u)}{s}\,\widehat{dG}(s)\,ds\right| \leqslant 2\gamma\,\text{I}\{|u| \leqslant 1+\gamma\}$ $\max_{|t|\leqslant 1}|\hat{p}'(t)|$ for all real u.

Note that the first integral in (4.5.20) is a convolution. One can also integrate by parts to represent $\widehat{pG}$ as a convolution-smoothing of the derivative $\hat{p}'$:

$$\widehat{pG} = \frac{i}{2\pi}\,\hat{p}' * \widetilde{dG}, \tag{4.5.22}$$

where

$$\widetilde{dG}(t) := \text{p.v.}\int_{-\infty}^{t} \widehat{dG}(s)\frac{ds}{s} := \lim_{\varepsilon\downarrow 0}\int_{(-\infty,t)\setminus(-\varepsilon,\varepsilon)} \widehat{dG}(s)\frac{ds}{s}$$

$$= \int_{-\gamma}^{-|t|} \widehat{dG}(s)\frac{ds}{s}\,\text{I}\{|t| < \gamma\} \tag{4.5.23}$$

for all real t. This follows because for all real u

$$\int_{\mathbb{R}} \frac{\hat{p}(u-s) - \hat{p}(u)}{s}\,\widehat{dG}(s)\,ds = \int_{\mathbb{R}} \widehat{dG}(s)\frac{ds}{s}\left(\int_{u}^{u-s} \hat{p}'(v)\,dv\,\text{I}\{s < 0\}\right.$$

$$\left. - \int_{u-s}^{u} \hat{p}'(v)\,dv\,\text{I}\{s > 0\}\right)$$

$$= \int_{u}^{\infty} \hat{p}'(v)\,dv\int_{-\infty}^{u-v} \widehat{dG}(s)\frac{ds}{s}$$

$$-\int_{-\infty}^{u}\hat{p}'(v)\,\mathrm{d}v\int_{u-v}^{\infty}\widehat{\mathrm{d}G}(s)\frac{\mathrm{d}s}{s}$$
$$=\int_{\mathbb{R}}\hat{p}'(v)\,\mathrm{d}v\,\widetilde{\widehat{\mathrm{d}G}}(u-v),$$

since the function $\widehat{\mathrm{d}G}$ is even. The latter condition or (4.5.23) also shows that the function $\widetilde{\widehat{\mathrm{d}G}}$ is even. Moreover, since $\widehat{\mathrm{d}G}(s)\to 1$ as $s\to 0$, (4.5.23) yields

$$\widetilde{\widehat{\mathrm{d}G}}(t)\sim \ln|t| \tag{4.5.24}$$

as $t\to 0$; thus, the function $\widetilde{\widehat{\mathrm{d}G}}$ is mildly singular in a neighborhood of 0. For instance, $\widetilde{\widehat{\mathrm{d}G}}(t)\equiv (1-|t|+\ln|t|)(\mathrm{I}\{|t|<1\}$ if $\widehat{\mathrm{d}G}(t)\equiv (1-|t|)$ $(\mathrm{I}\{|t|<1\}$.

Note also that in the case (prevented by the condition (4.5.11)) when $\widehat{\mathrm{d}G}=1$ on $\mathbb{R}$, the function $-2i\,\widehat{pG}$ would be the Hilbert transform of the function $\hat{p}$; see, for example, [48, Chapter V].

It follows from (4.5.19) and (4.5.21) that

$$M_2(t)=0 \quad \text{if } |t|>1+\gamma. \tag{4.5.25}$$

This condition on M_2 is obviously weaker than the condition

$$M_2(t)=0 \quad \text{if } |t|>1, \tag{4.5.26}$$

following from (4.2.1) and (4.2.5). However, by (4.5.19) and (4.5.9), one still has $M_1(t)=0$ if $|t|>1$, whereas the condition (4.5.26) was used in the *quick proof* beginning on page 116 only to bound two terms, $\mathfrak{G}_{2\alpha}(f_{31})$ and $\mathfrak{G}_{2\alpha}(f_{32})$. Therefore, one may expect the adverse impact of the weakening of the condition (4.5.26) to (4.5.25) to be rather limited and likely more than compensated for by the advantages provided by the more flexible construction of the smoothing filter M, with the tempered tilting of M_1. Moreover, the latter construction is, essentially, more general. Indeed, for instance, one may always include G into the scale family $(G_\alpha)_{\alpha>0}:=\left(G(\tfrac{\#}{\alpha})\right)_{\alpha>0}$, and then the tempered tilting (4.5.12) will be close to the harsh tilting (4.5.10) for large $\alpha>0$ provided that $G'(0)\neq 0$. Indeed, $G_\alpha(x)\sim \frac{G'(0)}{\alpha}x$ for each real $x\neq 0$ as $\alpha\to\infty$; of course, at that the value of $\kappa=\kappa_\alpha$ in (4.5.12) with $G=G_\alpha$ will be quite different from that in (4.5.10); in fact, the value of κ_* in (4.5.13) with $G=G_\alpha$ will then be asymptotically equivalent to the value of κ_* in (4.5.1) times $\frac{\alpha}{G'(0)}$, provided that $\int_{\mathbb{R}}|x|p(x)\,\mathrm{d}x<\infty$. At that, the value of γ in (4.5.11) for G will be replaced by the corresponding value $\gamma_\alpha := \frac{\gamma}{\alpha}$ for G_α, so that $\gamma_\alpha \to 0$ as $\alpha \to \infty$.

4.6 ANOTHER CONSTRUCTION OF SMOOTHING INEQUALITIES FOR NONUNIFORM BE BOUNDS

The construction described in this section may be more effective than the one introduced in Section 4.4.

Let us say that a function F is a scaled distribution function (scaled d.f., for brevity) if $F = \lambda F_0$ for some real $\lambda \geqslant 0$ and some d.f. F_0. For any scaled d.f. F, one can rewrite (4.2.2) in the following formally more general way: again for any real $T > 0$ and any real x,

$$\frac{1}{2} F(\infty-) + \mathfrak{G}\left(M_T(-\#)f_F(\#)\right)(x) \leqslant F(x-) \leqslant F(x+) \leqslant \frac{1}{2} F(\infty-)$$
$$+ \mathfrak{G}\left(M_T(\#)f_F(\#)\right)(x), \tag{4.6.1}$$

where f_F denotes the Fourier–Stieltjes transform $\int_{\mathbb{R}} e^{ix\#}\, dF(x)$ of F, and M_T and $\mathfrak{G}(f)$ are still as in (4.2.3) and (4.2.4).

Take any natural k and any r.v. X such that $\mathsf{E}|X|^k < \infty$, and introduce the functions $L_X, F_X, G_X, \hat{F}_X, \hat{G}_X$ defined by the formulas

$$L_X(x) := x^k\big(\mathsf{P}(X > x)\,\mathrm{I}\{x > 0\} - \mathsf{P}(X < x)\,\mathrm{I}\{x < 0\}\big),$$
$$F_X(x) := \mathsf{E}X^k\,\mathrm{I}\{X \leqslant x\}, \quad G_X(x) := \mathsf{E}(x_+ \wedge X)^k,$$
$$\hat{F}_X(t) := \mathsf{E}X^k e^{itX}, \qquad \hat{G}_X(t) := \int_0^1 k\alpha^{k-1}\mathsf{E}X^k e^{it\alpha X}\, d\alpha$$

for all real x and t; of course, the definition of the function L_X makes sense even without the condition $\mathsf{E}|X|^k < \infty$.

Note that

$$L_{-X}(\#) = (-1)^{k+1}L_X(-\#), \quad \hat{F}_{-X}(\#) = (-1)^k\hat{F}_X(-\#), \quad \text{and}$$
$$\hat{G}_{-X}(\#) = (-1)^k\hat{G}_X(-\#). \tag{4.6.2}$$

For a moment, consider the particular case when the r.v. X is nonnegative. Then

(i) $L_X = G_X - F_X$;
(ii) F_X and G_X are scaled d.f.'s, with $F_X(\infty-) = \mathsf{E}X^k = G_X(\infty-)$; also, G_X is continuous on $\mathbb{R}$;

(iii) $\hat{F}_X$ and $\hat{G}_X$ are the Fourier–Stieltjes transforms of F_X and G_X, respectively;

To check item (ii) on this list, use the dominated convergence. To verify item (iii) concerning $\hat{G}_X$ and G_X, note that $G_X(x) = \mu((-\infty, x])$ for all real x and $\int_\mathbb{R} e^{itx}\mu(\mathrm{d}x) = \hat{G}_X(t)$ for all real t, where μ is the nonnegative measure defined by the condition $\int_\mathbb{R} h\,\mathrm{d}\mu = \mathsf{E}\int_\mathbb{R} kz^{k-1}\mathrm{I}\{0 < z \leqslant X\}\, h(z)\,\mathrm{d}z$ for all bounded and/or nonnegative Borel functions $h\colon \mathbb{R} \to \mathbb{R}$; the relation between $\hat{F}_X$ and F_X is only easier to check.

Removing now the temporary assumption that the r.v. X is nonnegative and recalling (4.6.1), one has

$$\mathfrak{G}\left(M_T(-\#)\hat{G}_{X_+}(\#)\right)(x) - \mathfrak{G}\left(M_T(\#)\hat{F}_{X_+}(\#)\right)(x)$$
$$\leqslant L_{X_+}(x+) \leqslant L_{X_+}(x-) \tag{4.6.3}$$
$$\leqslant \mathfrak{G}\left(M_T(\#)\hat{G}_{X_+}(\#)\right)(x) - \mathfrak{G}\left(M_T(-\#)\hat{F}_{X_+}(\#)\right)(x) \tag{4.6.4}$$

for all real x. Using these inequalities (with $-X$ and $-x$ in place of X and x) together with the parity properties (4.6.2) and $\mathfrak{G}\left(f(-\#)\right)(-x) = -\mathfrak{G}(f)(x)$, one obtains the "negative" counterpart of the upper bound in (4.6.4):

$$L_{X_-}(x+) = (-1)^{k+1}L_{(-X)_+}((-x)-)$$
$$\leqslant (-1)^{k+1}\big[\mathfrak{G}\big(M_T((-1)^{k+1}\#)\hat{G}_{(-X)_+}(\#)\big)(-x)$$
$$- \mathfrak{G}\big(M_T((-1)^k\#)\hat{F}_{(-X)_+}(\#)\big)(-x)\big]$$
$$= \mathfrak{G}\big(M_T((-1)^k\#)\hat{G}_{X_-}(\#)\big)(x)$$
$$- \mathfrak{G}\big(M_T(-(-1)^k\#)\hat{F}_{X_-}(\#)\big)(x). \tag{4.6.5}$$

To proceed further, suppose that for some real constant $\varkappa$

$$\text{the function } (M(\#) - \varkappa)/\#\text{is in } L^1([-1,1]); \tag{4.6.6}$$

this condition was assumed in [43] and will be satisfied in the applications, usually with $\varkappa = 1$; it is even unclear whether the conditions (4.6.1), (4.2.1), and (4.2.5) can ever all hold without (4.6.6). Introducing the functions

$$M_{j,T}(\#) := M_j(\#/T)$$

for $j \in \{1,2\}$ (cf. (4.2.3) and (4.2.5)) and using (4.6.6) and the fact that $\left|\int_\varepsilon^A \frac{\sin zu}{u}\,\mathrm{d}u\right|$ is bounded uniformly over all real z and all ε and A such that $0 < \varepsilon < A$, one can easily show (cf. [43, (5)]) that the limit $\mathfrak{G}(M_{j,T}f)(x)$

exists (and is) in $\mathbb{C}$ for any $j \in \{1, 2\}$, any real $T > 0$, any characteristic function (c.f.) f, and any real x. This allows one to recombine the terms in the upper bounds in (4.6.4) and (4.6.5) to see that for any real $x \geqslant 0$

$$
\begin{aligned}
x^k \mathsf{P}(X \geqslant x) &= L_{X_+}(x-) = L_{X_+}(x-) + L_{X_-}(x+) \\
&\leqslant \mathfrak{G}\big(M_T(\#)\hat{G}_{X_+}(\#)\big)(x) - \mathfrak{G}\big(M_T(-\#)\hat{F}_{X_+}(\#)\big)(x) \\
&\quad + \mathfrak{G}\big(M_T((-1)^k\#)\hat{G}_{X_-}(\#)\big)(x) - \mathfrak{G}\big(M_T(-(-1)^k\#)\hat{F}_{X_-}(\#)\big)(x) \\
&= \mathfrak{G}\big(M_{1,T}\big[\hat{G}_{X_+} + \hat{G}_{X_-} - \hat{F}_{X_+} - \hat{F}_{X_-}\big]\big)(x) \\
&\quad + i\,\mathfrak{G}\big(M_{2,T}\big[\hat{G}_{X_+} + (-1)^k\hat{G}_{X_-} + \hat{F}_{X_+} + (-1)^k\hat{F}_{X_-}\big]\big)(x) \\
&= \mathfrak{G}\big(M_{1,T}\mathsf{E}X^k(W_X - V_X)\big)(x) + i\,\mathfrak{G}\big(M_{2,T}\mathsf{E}|X|^k(W_X + V_X)\big)(x),
\end{aligned}
\tag{4.6.7}
$$

where

$$
V_X(\#) := e^{iX\#} \quad \text{and} \quad W_X(\#) := \int_0^1 V_X(\alpha\#)k\alpha^{k-1}\,d\alpha
$$

$$
= \int_0^1 e^{i\alpha X\#}k\alpha^{k-1}\,d\alpha;
\tag{4.6.8}
$$

here we also used the obvious identities $\hat{F}_{X_\pm}(\#) = \mathsf{E}X_\pm^k e^{iX\#}$, $\hat{G}_{X_\pm}(\#) = \int_0^1 k\alpha^{k-1}\mathsf{E}X_\pm^k e^{i\alpha X\#}\,d\alpha$, $X_+^k + X_-^k = X^k$, and $X_+^k + (-X_-)^k = |X|^k$.

Quite similarly to the upper bound on $x^k \mathsf{P}(X \geqslant x)$ in (4.6.7), one can derive the corresponding lower bound on $x^k \mathsf{P}(X > x)$, with $-M_{2,T}$ in place of $M_{2,T}$. Thus, one obtains

Theorem 4.6.1. *Let M be any function such that the condition (4.6.1) holds for all d.f.'s F, all real x, and all real $T > 0$, as well as the conditions (4.2.1), (4.2.5), and (4.6.6). Then for all real $x \geqslant 0$ and all r.v.'s X such that $\mathsf{E}|X|^k < \infty$*

$$
\left| x^k \mathsf{P}(X \geqslant x) - \mathfrak{G}\left(M_{1,T}\,\mathsf{E}X^k(W_X - V_X)\right)(x)\right|
$$

$$
\leqslant i\,\mathfrak{G}\left(M_{2,T}\,\mathsf{E}|X|^k(W_X + V_X)\right)(x),
\tag{4.6.9}
$$

where W_X and V_X are as in (4.6.8); inequality (4.6.9) also holds with $\mathsf{P}(X > x)$ in place of $\mathsf{P}(X \geqslant x)$.

Remark 4.6.2. Introducing the c.f. of X,

$$
f(\#) := \mathsf{E}e^{iX\#},
$$

and its kth derivative $f^{(k)}$, one has $\mathsf{E}X^k V_X = i^{-k} f^{(k)}$ and $(\mathsf{E}X^k W_X)(\#) = i^{-k} \int_0^1 f^{(k)}(\alpha\#) k\alpha^{k-1}\, \mathrm{d}\alpha$, whence, concerning the left-hand side of (4.6.9),

$$\mathsf{E}X^k(W_X - V_X)(\#) = i^{-k} \int_0^1 [f^{(k)}(\alpha\#) - f^{(k)}(\#)] k\alpha^{k-1}\, \mathrm{d}\alpha. \quad (4.6.10)$$

Somewhat unfortunately, when k is odd the expression of the function $\mathsf{E}|X|^k(W_X + V_X)$ in the right-hand side of (4.6.9) in terms of the c.f. f (cf., e.g., [34]) is much less convenient than the expression for $\mathsf{E}X^k(W_X - V_X)$ in (4.6.10)—and the case most interesting in the applications is that of $k = 3$.

However, there is a simple and apparently rather effective way to deal with this inconvenience:

Proposition 4.6.3. *Under the conditions of Theorem 4.6.1,*

$$\left| \mathfrak{G}\big(M_{2,T}\, \mathsf{E}|X|^k(W_X + V_X)\big)(x) \right.$$
$$\left. -i^{-k} \int_0^1 k\alpha^{k-1}\, \mathfrak{G}\big(M_{2,T}(\#)\,[f^{(k)}(\alpha\#) + f^{(k)}(\#)]\big)(x)\, \mathrm{d}\alpha \right|$$
$$\leqslant \frac{c_{2,p}}{\pi} \frac{2k-p}{k-p}\, \mathsf{E}\frac{|X_-|^k}{(|X_-|+x)^p}\frac{1}{T^p}$$
$$\leqslant \frac{c_{2,p}}{\pi} \frac{2k-p}{k-p}\left(\mathsf{E}|X_-|^{k-p} \wedge \frac{\mathsf{E}|X_-|^k}{x^p} \right)\frac{1}{T^p} \qquad (4.6.11)$$

for all $p \in (0, k)$ and all real $x > 0$, where

$$c_{2,p} := \sup_{u \in \mathbb{R}} |u|^p |\widehat{N_2}(u)|$$

and $\widehat{N_2}$ is the Fourier transform of the function $N_2(\#) := M_2(\#)/\#$.

Proof of Proposition 4.6.3. Note that the left-hand side of the first inequality in (4.6.11) equals

$$\mathrm{LHS} := \left| \mathfrak{G}\big(M_{2,T}\, \mathsf{E}|X|^k(W_X + V_X)\big)(x) - \mathfrak{G}\big(M_{2,T}\, \mathsf{E}X^k(W_X + V_X)\big)(x) \right|$$
$$= 2\left| \mathfrak{G}\big(M_{2,T}\, \mathsf{E}X_-^k(W_X + V_X)\big)(x) \right|;$$

cf. (4.6.10). Further, by (4.2.4),

$$\mathrm{LHS} = \frac{1}{\pi}\left| \int_{\mathbb{R}} e^{-itx} \frac{M_2(t/T)}{t}\, \mathsf{E}X_-^k \left(\int_0^1 k\alpha^{k-1} e^{i\alpha tX}\, \mathrm{d}\alpha + e^{itX} \right) \mathrm{d}t \right|$$
$$= \frac{1}{\pi}\left| \mathsf{E}X_-^k \int_{\mathbb{R}} N_2(u) \left(\int_0^1 k\alpha^{k-1} e^{i\alpha Tu(X-x)}\, \mathrm{d}\alpha + e^{iTu(X-x)} \right) \mathrm{d}u \right|$$

$$= \frac{1}{\pi} \left| \mathsf{E}|X_-|^k \left(\int_0^1 k\alpha^{k-1} \widehat{N_2}\left(\alpha T(X-x)\right) \, d\alpha + \widehat{N_2}\left(T(X-x)\right) \right) du \right|$$

$$\leqslant \frac{c_{2,p}}{\pi} \mathsf{E} \frac{|X_-|^k}{T^p |X-x|^p} \left(\int_0^1 k\alpha^{k-1-p} \, d\alpha + 1 \right)$$

$$= \frac{c_{2,p}}{\pi} \mathsf{E} \frac{|X_-|^k}{(|X_-|+x)^p} \frac{2k-p}{k-p} \frac{1}{T^p};$$

here we used the equality $|X-x| = |X_-| + x$, valid for any real $x > 0$ on the event $\{X_- \neq 0\}$. Thus, the first inequality in (4.6.11) is verified, and the second inequality there follows because $|X_-| + x \geqslant |X_-| \vee x$ for $x > 0$. $\quad\square$

Remark 4.6.4. For instance, for Prawitz's particular function M as in (4.2.7), $N_2(\#) = -\pi(1 - |\#|)_+$, so that $\widehat{N_2}(x) = -\pi \left(\frac{\sin x/2}{x/2} \right)^2$ for real $x \neq 0$, whence $c_{2,2} = 4\pi$ and $c_{2,1} = 4\pi \sup_{x>0} \frac{\sin^2 x/2}{x}$. It follows that for $k = 3$ the second upper bound in (4.6.11) is $16 \left(\mathsf{E}|X_-| \wedge \frac{\mathsf{E}|X_-|^3}{x^2} \right) \frac{1}{T^2}$ if p is taken to be 2, and it is no greater than $3.6231 \left(\mathsf{E}|X_-|^2 \wedge \frac{\mathsf{E}|X_-|^3}{x} \right) \frac{1}{T}$ with $p = 1$. Here one can obviously further bound the moments of $|X_-|$ from above by the corresponding moments of $|X|$; these bounds can be obviously improved if the distribution of X is symmetric.

For simplicity, let us consider here the iid case and accordingly let $X := S/\sqrt{n}$, so that X is a zero-mean unit-variance r.v. Then, with $k = 3$ and $p = 2$,

$$\mathsf{E} \frac{|X_-|^3}{(|X_-|+x)^2} \leqslant \mathsf{E}|X_-| \wedge \frac{\mathsf{E}|X_-|^3}{x^2} \leqslant 1 \wedge \frac{\mathsf{E}|X|^3}{x^2} \leqslant 1 \wedge \frac{2 + \beta_3/\sqrt{n}}{x^2} \tag{4.6.12}$$

by the Rosenthal-type inequality (see, e.g., [4, Lemma 6.3] or [41, (12)])

$$\mathsf{E}|X|^3 \leqslant 2 + \beta_3/\sqrt{n}, \tag{4.6.13}$$

where $\beta_3 := \mathsf{E}|X_1|^3$; this may be compared with $\mathsf{E}|Z|^3 = 2\sqrt{\frac{2}{\pi}} \approx 1.6$, where $Z \sim N(0,1)$. In view of Markov's inequality and the mentioned value 0.4748 of c_{u}, (4.6.13) also yields (4.3.1) for all real $x \geqslant 0$ with $c_{\mathrm{nu}} = 4.5$ in the "small n" case when $\frac{\beta_3}{\sqrt{n}} \geqslant \frac{2}{3}$. So, if one recalls (Subsection 4.3.1) that the apparently best known upper bound on c_{nu} in the iid case is over 17 and thus considers the value 4.5 for c_{nu} satisfactory at this point, then without loss of generality (w.l.o.g.) $\frac{\beta_3}{\sqrt{n}} < \frac{2}{3}$. Moreover, comparing the desired nonuniform bound $4.5\frac{\beta_3}{(1+x^3)\sqrt{n}}$ with the known uniform bound $0.4748\frac{\beta_3}{\sqrt{n}}$, one sees that w.l.o.g. $x > x_0 := (\frac{4.5}{0.4748} - 1)^{1/3} = 2.039\ldots$

Another upper bound on the term $\mathsf{E}\frac{|X_-|^k}{(|X_-|+x)^p}$, which is apparently better than the upper bound in (4.6.12), is as follows. Again, let us consider the case of principal interest, when $k = 3$. At that, to be specific, take $p = 2$.

Consider the function $h(\#) := h_x(\#) := \frac{|\#_-|^3}{(|\#_-|+x)^2}$, for any given real $x > 0$. Then it is easy to see that $|h'''(u)| \leqslant \frac{6}{x^2}$ for all real nonzero u. Hence, by Tyurin's result [51, Theorem 2],

$$\mathsf{E}\frac{|X_-|^3}{(|X_-|+x)^2} \leqslant \frac{\psi(x) + \beta_3/\sqrt{n}}{x^2}, \qquad (4.6.14)$$

where $\psi(x) := x^2\mathsf{E}\frac{|Z_-|^3}{(|Z_-|+x)^2}$, so that the function ψ is increasing on the interval $(0, \infty)$, from $\psi(0+) = 0$ to $\psi(\infty-) = \mathsf{E}|Z_-|^3 = \sqrt{\frac{2}{\pi}} = 0.797\ldots < 0.8$.

Thus, the upper bound in (4.6.14) is less than $\frac{0.8+2/3}{x_0^2} < 0.36 < 1$ and hence indeed significantly less than the upper bound in (4.6.12). Note also that the increase of ψ is rather slow; in particular, $\psi(3.5) \approx 0.35$, whereas certain considerations show that the most "difficult" values of x are between $x_0 \approx 2$ and about 3.5.

One can also try to use the more accurate upper bound

$$\frac{1}{\alpha^p T^p ||X_-| + x|^p} \sup\{u^p|\widehat{N_2}(u)| : u \geqslant \alpha T x\} \qquad (4.6.15)$$

on $\widehat{N_2}(\alpha T(|X_-| + x))$—instead of the bound $\frac{1}{\alpha^p T^p ||X_-|+x|^p} \sup\{|u|^p |\widehat{N_2}(u)| : u \in \mathbb{R}\}$, essentially used in the proof of Proposition 4.6.3. At that, one may want to utilize a function M with its imaginary part M_2 smoother than that of the Prawitz particular function, so that the Fourier transform $\widehat{N_2}$ of the function $N_2(\#) := M_2(\#)/\#$ be decreasing faster.

In Section 4.4, a quick proof of (4.3.1) was given. Using Theorem 4.6.1 in this section (with $k = 3$), we can now give the following, yet quicker proof of (4.3.1), in which we have fewer terms to bound than in the "quick proof" in Section 4.4.

A quicker proof of Nagaev's nonuniform BE bound (4.3.1)

Let $T = c_T \sqrt{n}/\beta_3$, where c_T is a small enough positive real constant. Let $A \lesssim B$ mean $|A| \leq CB$ for some absolute constant C. Let $X := S/\sqrt{n}$. If $T \leq 1$ then $1 \leq \frac{\beta_3}{\sqrt{n}}$. So, for all real $x \geq 0$, by the Markov and Rosenthal inequalities, $(1+x^3)\mathsf{P}(X \geq x) \leq 1 + \mathsf{E}|X|^3 \leq 1 + \frac{\beta_3}{\sqrt{n}} \lesssim \frac{\beta_3}{\sqrt{n}}$ and similarly $(1 + x^3)\mathsf{P}(Z \geq x) \lesssim \frac{\beta_3}{\sqrt{n}}$, whence (4.3.1) follows.

It remains to consider the case $T > 1$. Note that then $n > (\beta_3/c_T)^2 \geq 3$ and hence $n \geq 4$ provided that $c_T \leq 1/\sqrt{3}$.

In view of the uniform BE bound, Theorem 4.6.1 (with M as in (4.2.7), say), (4.6.10), Proposition 4.6.3, and Remark 4.6.4, in order to prove (4.3.1) it is enough to show that $\mathfrak{G}_{1\alpha}(f''' - g''') \lesssim \frac{\beta_3}{\sqrt{n}}$ and $\mathfrak{G}_{2\alpha}(f''') \lesssim \frac{\beta_3}{\sqrt{n}}$ for $\alpha \in (0,1]$, where f is the c.f. of $X := S/\sqrt{n}$, $g(\#) := e^{-\#^2/2}$, and

$$\mathfrak{G}_{j\alpha}(h)(x) := \mathfrak{G}\left(M_j\left(\frac{\alpha\#}{T}\right) h(\alpha\#)\right)(x).$$

For $j \in \{0, 1, 2, 3\}$, introduce $f_1^{(j)}(t) := \left(\frac{\mathrm{d}}{\mathrm{d}t}\right)^j f_1(t)$ and $f_{1n}^{(j)}(t) := f_1^{(j)}(t/\sqrt{n})$, where f_1 denotes the c.f. of X_1.

Similarly, starting with $g_1 := g$ in place of f_1, define $g_{1n}^{(j)}$, and then let $d_{1n}^{(j)} := f_{1n}^{(j)} - g_{1n}^{(j)}$ and $h_{1n}^{[j]} := \left|f_{1n}^{(j)}\right| \vee \left|g_{1n}^{(j)}\right|$; omit superscripts $^{(0)}$ and $^{[0]}$. Note that $f = f_{1n}^n$ and hence $\sqrt{n}f''' = f_{31} + f_{32} + f_{33}$, where $f_{31} := (n-1)(n-2)f_{1n}^{n-3}\left(f_{1n}^{(1)}\right)^3$, $f_{32} := 3(n-1)f_{1n}^{n-2}f_{1n}^{(1)}f_{1n}^{(2)}$, and $f_{33} := f_{1n}^{n-1}f_{1n}^{(3)}$; do similarly with g and g_1 in place of f and f_1. By Remark 4.2.1 and Proposition 4.4.3, $M_j\left(\frac{\alpha\#}{T}\right)f_{33}/\beta_3$ is a quasi-c.f. and hence, by Proposition 4.4.2, $\mathfrak{G}_{j\alpha}(f_{33}) \lesssim \beta_3$, for $j \in \{1, 2\}$.

So, it suffices to show that $\mathfrak{G}_{1\alpha}(f_{3k} - g_{3k}) \lesssim \beta_3$ and $\mathfrak{G}_{2\alpha}(f_{3k}) \lesssim \beta_3$ for $k \in \{1, 2\}$.

This can be done in a straightforward manner using the following estimates for $j \in \{0, 1, 2, 3\}$ and $|t| \leq T$: $M_1 \leq 1$, $M_2(\frac{t}{T}) \leq \frac{|t|}{T} \lesssim |t|\beta_3/\sqrt{n}$, $h_{1n}(t)^{n-j} \leq e^{-ct^2}$ (where c is a positive real number depending only on the choice of c_T), $h_{1n}^{[1]}(t) \leq |t|/\sqrt{n}$, $h_{1n}^{[2]}(t) \leq 1$, $|d_{1n}^{(j)}(t)| \leq$

$\beta_3(|t|/\sqrt{n})^{3-j}$, and hence $f_{1n}^{n-j}(t) - g_{1n}^{n-j}(t) \lesssim |t|^3 e^{-ct^2}\beta_3/\sqrt{n}$; cf., for example, [28, Ch. V, Lemma 1]. For instance, $|f_{31}-g_{31}| \lesssim n^2(D_{311}+D_{312})$,

where $D_{311}(t) := \left(|f_{1n}^{n-3} - g_{1n}^{n-3}| \left(h_{1n}^{[1]} \right)^3 \right)(t) \lesssim |t|^3 e^{-ct^2} \frac{\beta_3}{\sqrt{n}} \left(\frac{|t|}{\sqrt{n}} \right)^3$ and

$D_{312}(t) := \left(h_{1n}^{n-3} \left(h_{1n}^{[1]} \right)^2 |d_{1n}^{(1)}| \right)(t) \lesssim e^{-ct^2} \left(\frac{|t|}{\sqrt{n}} \right)^2 \beta_3 \left(\frac{|t|}{\sqrt{n}} \right)^2$, so that

$\mathfrak{G}_{1\alpha}(f_{31} - g_{31}) \lesssim \int_{-\infty}^{\infty}(t^6 + t^4)e^{-ct^2}\beta_3 \frac{dt}{|t|} \lesssim \beta_3$.

REFERENCES

[1] V. Bentkus, On the asymptotical behavior of the constant in the Berry-Esseen inequality, J. Theoret. Probab. 7 (2) (1994) 211–224.

[2] A. Bikelis, Estimates of the remainder term in the central limit theorem, Litovsk. Mat. Sb. 6 (1966) 323–346.

[3] H. Bohman, To compute the distribution function when the characteristic function is known, Skand. Aktuarietidskr. 1963 (1964) 41–46.

[4] L.H.Y. Chen, Q.-M. Shao, Stein's method for normal approximation, in: An Introduction to Stein's Method, Lect. Notes Ser. Inst. Math. Sci. Natl. Univ. Singap., vol. 4, Singapore University Press, Singapore, 2005, pp. 1–59.

[5] G.P. Chistyakov, On a problem of A. N. Kolmogorov, Zap. Nauchn. Sem. Leningrad. Otdel. Mat. Inst. Steklov. (LOMI) 184 (Issled. po Mat. Statist. 9) (1990) 289–319, 326–327.

[6] T. Dinev, L. Mattner, The Asymptotic Berry-Esseen Constant for Intervals, 2011, http://arxiv.org/abs/1111.7146.

[7] W.F. Jr. Donoghue, Distributions and Fourier transforms, Pure and Applied Mathematics, vol. 32, Academic Press, New York, 1969.

[8] C.G. Esseen, A moment inequality with an application to the central limit theorem, Skand. Aktuarietidskr. 39 (1956) 160–170.

[9] S.V. Gavrilenko, Refinement of nonuniform estimates of the rate of convergence of the distributions of Poisson random sums to the normal law, Inform. Primen. (Inform. Appl.) 5 (2011) 12–24 (in Russian).

[10] M.E. Grigor'eva, S.V. Popov, On nonuniform estimates of the rate of convergence in the central limit theorem, Sistemy i Sredstva Inform. 22 (1) (2012) 180204, http://mi.mathnet.ru/ssi274.

[11] M.E. Grigor'eva, S.V. Popov, An upper bound for the absolute constant in the nonuniform version of the Berry-Esseen inequalities for nonidentically distributed summands, Dokl. Math. 445 (4) (2012) 380382, http://link.springer.com/article/10.1134%2FS1064562412040242.

[12] J. Gurland, Inversion formulae for the distribution of ratios, Ann. Math. Stat. 19 (1948) 228–237.

[13] W. Hoeffding, The extrema of the expected value of a function of independent random variables, Ann. Math. Stat. 26 (1955) 268–275.

[14] A.F. Karr, Extreme points of certain sets of probability measures, with applications, Math. Oper. Res. 8 (1) (1983) 74–85.

[15] V.Y. Korolev, I.G. Shevtsova, An improvement of the Berry–Esseen inequality with applications to Poisson and mixed Poisson random sums, Scand. Actuar. J. 2012 (2) (2012) 81–105.

[16] E. Lukacs, Characteristic Functions, second ed. (revised and enlarged), Hafner Publishing Co., New York, 1970.

[17] R. Michel, On the constant in the nonuniform version of the Berry-Esseen theorem, Z. Wahrsch. Verw. Gebiete 55 (1) (1981) 109–117.

[18] S.A. Mirakhmedov, On the absolute constant in a nonuniform estimate of the rate of convergence in the central limit theorem, Izv. Akad. Nauk UzSSR Ser. Fiz.-Mat. Nauk (4) (1984) 26–30.

[19] A.V. Nagaev, Integral limit theorems with regard to large deviations when Cramér's condition is not satisfied. I, Theor. Probab. Appl. 14 (1969) 51–64.

[20] A.V. Nagaev, Integral limit theorems with regard to large deviations when Cramér's condition is not satisfied. II, Theor. Probab. Appl. 14 (1969) 193–208.

[21] S.V. Nagaev, Some limit theorems for large deviations, Teor. Verojatnost. i Primenen 10 (1965) 231–254.

[22] Y.S. Nefedova, I.G. Shevtsova, On the accuracy of normal approximation for the distributions of Poisson random sums, Inform. Primen. (Inform. Appl.) 5 (1) (2011) 39–45 (in Russian).

[23] Y.S. Nefedova, I.G. Shevtsova, Nonuniform estimates of convergence rate in the central limit theorem, Teor. Veroyatnost. i Primenen. 57 (1) (2012) 62–97 (in Russian).

[24] L. Paditts, S.A. Mirakhmedov, Letter to the editors: "On the absolute constant in a nonuniform estimate of the rate of convergence in the central limit theorem", [Izv.Akad.Nauk UzSSR Ser.Fiz.-Mat.Nauk 1984, no. 4, 26–30; MR MR0780094 (86i:60071)], Izv. Akad. Nauk UzSSR Ser. Fiz.-Mat. Nauk (3) (1986) 80.

[25] L. Paditz, Über die Annäherung der Verteilungsfunktionen von Summen unabhängiger Zufallsgröen gegen unbegrenzt teilbare Verteilungsfunktionen unter besonderer beachtung der Verteilungsfunktion der standardisierten Normalverteilung. Dissertation A, PhD thesis, Technische Universität Dresden, 1977.

[26] L. Paditz, Abschätzungen der Konvergenzgeschwindigkeit zur Normalverteilung unter Voraussetzung einseitiger Momente, Math. Nachr. 82 (1978) 131–156.

[27] L. Paditz, On the analytical structure of the constant in the nonuniform version of the Esseen inequality, Statistics 20 (3) (1989) 453–464.

[28] V.V. Petrov, Sums of Independent Random Variables, Springer-Verlag, New York, 1975, translated from the Russian by A. A. Brown, Ergebnisse der Mathematik und ihrer Grenzgebiete, Band 82.

[29] R.R. Phelps, Lectures on Choquet's Theorem, D. Van Nostrand Co., Inc., Princeton, NJ/Toronto, Ont./London, 1966.

[30] I. Pinelis, Improved nonuniform Berry–Esseen-type bounds, Preprint, http://arxiv.org/find/all/1/au:+pinelis/0/1/0/all/0/1.

[31] I. Pinelis, On the Bennett–Hoeffding inequality, 2009, a shorter version appeared in [37], http://arxiv.org/abs/0902.4058.

[32] I. Pinelis, Exact bounds on the truncated-tilted mean, with applications, Preprint, 2011, http://arxiv.org/abs/1103.3683.

[33] I. Pinelis, Exact lower bounds on the exponential moments of Winsorized and truncated random variables, J. Appl. Probab. 48 (2011) 547–560.

[34] I. Pinelis, Positive-part moments via the Fourier–Laplace transform, J. Theor. Probab. 24 (2011) 409–421.

[35] I. Pinelis, An exact bound on the truncated-tilted mean for symmetric distributions, 2012, http://arxiv.org/abs/1205.5234.

[36] I. Pinelis, Optimal re-centering bounds, with applications to Rosenthal-type concentration of measure inequalities, in: High Dimensional Probability VI (The Banff Volume), Progr. Probab., vol. 66, Birkhäuser, Basel, 2013, pp. 81–93, http://arxiv.org/abs/1111.2622.

[37] I. Pinelis, On the Bennett–Hoeffding inequality, Ann. Inst. Henri Poincaré Probab. Stat. 50 (1) (2014) 15–27.

[38] I. Pinelis, On the extreme points of moments sets, Math. Meth. Oper. Res. 83 (3) (2016) 325–349.

[39] I.F. Pinelis, A problem on large deviations in a space of trajectories, Theory Probab. Appl. 26 (1) (1981) 69–84.

[40] I.F. Pinelis, Asymptotic equivalence of the probabilities of large deviations for sums and maximum of independent random variables, in: Limit Theorems of Probability Theory, Trudy Inst. Mat., vol. 5, "Nauka" Sibirsk. Otdel, Novosibirsk, 1985, pp. 144–173.

[41] I.F. Pinelis, Rosenthal-type inequalities for martingales in 2-smooth Banach spaces, Theory Probab. Appl. 59 (4) (2015) 699–706.

[42] I.F. Pinelis, S.A. Utev, Sharp exponential estimates for sums of independent random variables, Theory Probab. Appl. 34 (2) (1989) 340–346.

[43] H. Prawitz, Limits for a distribution, if the characteristic function is given in a finite domain, Skand. Aktuarietidskr. 55 (1972) 138–154.

[44] I. Shevtsova, On the absolute constants in the Berry-Esseen type inequalities for identically distributed summands, 2011, http://arxiv.org/abs/1111.6554.

[45] I. Shevtsova, On the absolute constants in Nagaev-Bikelis-type inequalities, in: I. Pinelis (Ed.), Inequalities and Extremal Problems in Probability and Statistics: Selected Topics, Elsevier, Amsterdam, 2016.

[46] I.G. Shevtsova, Refinement of estimates for the rate of convergence in Lyapunov's theorem, Dokl. Akad. Nauk 435 (1) (2010) 26–28.

[47] I.S. Shiganov, Refinement of the upper bound of a constant in the remainder term of the central limit theorem, in: Stability Problems for Stochastic Models (Moscow, 1982), Vsesoyuz. Nauchno-Issled. Inst. Sistem. Issled., Moscow, 1982, pp. 109–115.

[48] E.C. Titchmarsh, Introduction to the Theory of Fourier Integrals, third ed., Chelsea Publishing Co., New York, 1948.

[49] W. Tysiak, Gleichmäige und nicht-gleichmige Berry-Esseen-Abschätzungen. Dissertation, PhD thesis, 1983.

[50] I. Tyurin, New estimates of the convergence rate in the Lyapunov theorem, 2009, http://arxiv.org/abs/0912.0726.

[51] I.S. Tyurin, Some optimal bounds in the central limit theorem using zero biasing, Stat. Probab. Lett. 82 (3) (2012) 514–518.

[52] J.D. Vaaler, Some extremal functions in Fourier analysis, Bull. Am. Math. Soc. 12 (2) (1985) 183–216.

[53] G. Winkler, Extreme points of moment sets, Math. Oper. Res. 13 (4) (1988) 581–587.

CHAPTER 5

On the Berry–Esseen Bound for the Student Statistic

Iosif Pinelis

Michigan Technological University, Houghton, MI, United States

5.1 SUMMARY AND DISCUSSION

Consider the self-normalized sum

$$T := \frac{S}{V},$$

where

$$S := \sum_1^n X_i, \quad V := \sqrt{\sum_1^n X_i^2},$$

and $X_1, \dots, X_n$ are independent zero-mean random variables (r.v.'s). It is assumed that $T = 0$ on the event $\{V = 0\}$. For any $p \in (0, \infty)$, introduce also

$$\beta_p := \sum_1^n \mathsf{E}\,|X_i|^p \quad \text{and} \quad \tilde{\beta}_p := \sum_1^n \mathsf{E}\,|X_i^2 - \mathsf{E}\,X_i^2|^{p/2},$$

assuming that $0 < \beta_3 < \infty$ (and hence $0 < \beta_2 < \infty$).

Let Φ be the standard normal distribution function.

Theorem 5.1.1. *One has*

$$|\mathsf{P}(T \leqslant z) - \Phi(z)| \leqslant A_3 \frac{\beta_3}{\beta_2^{3/2}} + A_4 \frac{\tilde{\beta}_4^{1/2}}{\beta_2} + A_6 \frac{\tilde{\beta}_6}{\beta_3^3 \beta_2^{3/2}} \quad (5.1.1)$$

for all $z \in \mathbb{R}$ and for all triples $\tau := (A_3, A_4, A_6)$ of absolute constants belonging to the set $\mathcal{T} := \{\tau_1, \tau_2, \tau_3\}$ of triples, where

$$\tau_1 := (2.02, 1.10, 0.15),$$
$$\tau_2 := (16.25, 2.27, 4.83 \times 10^{-6}),$$
$$\tau_3 := (1.37, 40.54, 43.65).$$

Inequalities and Extremal Problems in Probability and Statistics. http://dx.doi.org/10.1016/B978-0-12-809818-9.00005-7

The triple $\tau_1 = (2.02, 1.10, 0.15)$ of the constant factors A_3, A_4, A_6 was obtained trying to minimize the sum $A_3 + A_4 + A_6$ of the constants; for details, see the proof (in Section 5.2) of Theorem 5.1.1 and especially the table at the end of that proof. The triple τ_2 was obtained trying to minimize the effect of the sixth-order moments of the X_i's. The triple τ_3 was designed to work best when $\tilde{\beta}_4$ and $\tilde{\beta}_6$ are very small, that is, when the distribution of each X_i is close to the symmetric distribution on a symmetric two-point set. The triples τ_1, τ_2, τ_3 will be used in this paper to compare the upper bound in (5.1.1) with one due to Shao [19].

In the i.i.d. case, that is, when the r.v.'s $X_1, \ldots, X_n$ are independent copies of an r.v. X, one can improve the values A_3, A_4, A_6 of the absolute constants in (5.1.1); in this case, let us assume without loss of generality that

$$\mathsf{E}\,X^2 = 1.$$

Introduce

$$\rho_3 := \mathsf{E}\,|X|^3, \quad \rho_4 := \sqrt{\mathsf{E}(X^2 - 1)^2}, \quad \rho_6 := \frac{\mathsf{E}\,|X^2 - 1|^3}{\mathsf{E}\,|X|^3}. \qquad (5.1.2)$$

Theorem 5.1.2. *If $X, X_1, \ldots, X_n$ are i.i.d. r.v.'s with $\mathsf{E}\,X = 0$, $\mathsf{E}\,X^2 = 1$, and $\mathsf{E}\,|X|^3 < \infty$, then*

$$|\mathsf{P}(T \leqslant z) - \Phi(z)| \leqslant \frac{A_3\rho_3 + A_4\rho_4 + A_6\rho_6}{\sqrt{n}} \qquad (5.1.3)$$

for all $z \in \mathbb{R}$ and for all triples $\tau := (A_3, A_4, A_6)$ of absolute constants belonging to the set $\tilde{\mathcal{T}} := \{\tilde{\tau}_1, \tilde{\tau}_2, \tilde{\tau}_3, \tilde{\tau}_{1.1}\}$ of triples, where

$$\tilde{\tau}_1 := (1.93, 1.06, 0.14),$$
$$\tilde{\tau}_2 := (15.54, 2.11, 4.72 \times 10^{-6}),$$
$$\tilde{\tau}_3 := (1.27, 50.15, 54.10),$$
$$\tilde{\tau}_{1.1} := (2.35, 0.99, 0.11).$$

For each $i = 1, 2, 3$, the triple $\tilde{\tau}_i$ in Theorem 5.1.2 is to be compared with the triple τ_i in Theorem 5.1.1; the triple $\tilde{\tau}_{1.1}$ is a modification of $\tilde{\tau}_1$, which will be used in a comparison with the bound in (5.1.9).

For $n \geqslant 2$, the Student statistic

$$t := \frac{\overline{X}\sqrt{n}}{\sqrt{\frac{1}{n-1}\sum_1^n (X_i - \overline{X})^2}},$$

where $\overline{X} := \frac{1}{n}\sum_1^n X_i$, can be expressed as a monotonic transformation of the self-normalized sum T:

$$t = \sqrt{\frac{n-1}{n}} \frac{T}{\sqrt{1-T^2/n}}. \tag{5.1.4}$$

Therefore, one immediately has

Corollary 5.1.3. *Theorems 5.1.1 and 5.1.2 hold if* $\mathsf{P}(T \leqslant z) - \Phi(z)$ *is replaced there by* $\mathsf{P}(t \leqslant z) - \Phi_n(z)$, *where*

$$\Phi_n(z) := \Phi\left(\frac{z}{\sqrt{1+(z^2-1)/n}}\right). \tag{5.1.5}$$

A Berry–Esseen type of bound of the optimal order for the Student statistic of i.i.d. X_i's was obtained in 1996 by Bentkus and Götze [2], using a Fourier transformation method. This was extended to the non-i.i.d. case by Bentkus et al. [1], whose result can be rewritten as follows:

$$\left| \mathsf{P}(t \leqslant z) - \Phi\left(z\sqrt{\tfrac{n}{n-1}}\right) \right| \leqslant C_2\gamma_2 + C_3\gamma_3, \tag{5.1.6}$$

where C_2 and C_3 are absolute constants,

$$\gamma_2 := \frac{1}{\beta_2}\sum_1^n \mathsf{E}\, X_i^2\, \mathbf{I}\left\{|X_i| > \frac{\sqrt{\beta_2}}{2}\right\},$$

$$\gamma_3 := \frac{1}{\beta_2^{3/2}}\sum_1^n \mathsf{E}\,|X_i|^3\, \mathbf{I}\left\{|X_i| \leqslant \frac{\sqrt{\beta_2}}{2}\right\}. \tag{5.1.7}$$

Note that $t \sim T$ as $n \to \infty$. The function Φ_n, defined by (5.1.5), may be considered as an improper distribution function, with the "impropriety" $1-\big(\Phi_n(\infty)-\Phi_n(-\infty)\big) = 2\big(1-\Phi(\sqrt{n})\big) \sim \sqrt{\frac{2}{\pi n}}\, e^{-n/2}$ for large n, which is much less than $\frac{1}{\sqrt{n}}$. If n is not very large, the tail probability $1-\Phi_n(z)$ may be much greater than $1 - \Phi(z)$, which appears to correspond qualitatively to the fact that the tail of the Student distribution is significantly heavier than the standard normal tail when the number of degrees of freedom (d.f.) is not large. This heuristics appears to be confirmed by Fig. 5.1, for $n = 10$; the pictures for $n = 5$ and $n = 20$ look quite similarly.

It appears that on the interval $[1.5, \infty)$ the tail function $1 - \Phi_n(\cdot)$ is closer to that of the Student distribution than the tail functions $1 - \Phi(\cdot)$ and $1-\Phi(\cdot\sqrt{\frac{n}{n-1}})$ are. So, while the method of the proof (given in Section 5.2) appears to allow one to obtain analogs of Theorems 5.1.1 and 5.1.2 for

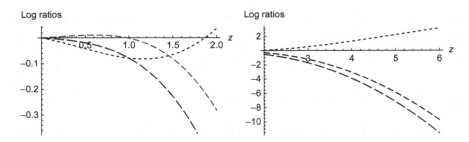

Fig. 5.1 *Logarithms of the ratios of the tail functions* $1 - \Phi(\cdot)$ *(longest dashes),* $1 - \Phi_n(\cdot)$ *(medium dashes), and* $1 - \Phi(\cdot \sqrt{\frac{n}{n-1}})$ *(shortest dashes) to the tail function of the Student distribution with* $n - 1$ *d.f.*

$\mathsf{P}(t \leqslant z) - \Phi(z)$ in place of $\mathsf{P}(T \leqslant z) - \Phi(z)$ or $\mathsf{P}(t \leqslant z) - \Phi_n(z)$, such analogs will not be pursued here.

Anyway, the following proposition shows that $\Phi_n(z)$ differs from $\Phi(z)$ by much less than $1/\sqrt{n}$, uniformly in $z \in \mathbb{R}$.

Proposition 5.1.4. *For all* $n > 1$ *and* $z \in \mathbb{R}$ *we have*

$$|\Phi(z) - \Phi_n(z)| < \frac{C}{n - 1}, \quad where \tag{5.1.8}$$

$$C := \left(k - \frac{1}{2}\right)e^{-k}\sqrt{\frac{k}{\pi}} = 0.162\ldots \quad and \quad k := 1 + \frac{\sqrt{3}}{2};$$

this constant factor, C, *is the best possible in* (5.1.8).

One may be concerned that it is more natural to compare the distribution function of the statistic t (as in (5.1.4), for general zero-mean X_i's), not with Φ or Φ_n, but with the distribution function (say F_{n-1}) of Student's distribution with $n - 1$ d.f.—that is, with the distribution function of the statistic t for i.i.d. standard normal X_i's. However, as shown in [15],

$$|F_{n-1}(z) - \Phi(z)| < \frac{\tilde{C}}{n - 1} \quad \text{with} \quad \tilde{C} = 0.158\ldots$$

for all $n \geqslant 5$ and $z \in \mathbb{R}$. Therefore and in view of Proposition 5.1.4, $F_{n-1}(z)$ differs from $\Phi_n(z)$ by much less than $1/\sqrt{n}$, uniformly in $z \in \mathbb{R}$. Thus, Corollary 5.1.3 is quite relevant, notwithstanding the mentioned concern.

In the i.i.d. case, Nagaev [7, 1.18] stated an inequality, which reads as follows (in the conditions of Theorem 5.1.2): for all $z \in \mathbb{R}$,

$$|\mathsf{P}(T \leqslant z) - \Phi(z)| < \left(4.4\,\mathsf{E}\,|X|^3 + \frac{\mathsf{E}\,X^4}{\mathsf{E}\,|X|^3} + \mathsf{E}\,|X^2 - 1|^3\right)\frac{1}{\sqrt{n}}. \quad (5.1.9)$$

Clearly, $\mathsf{E}\,X^2 = 1$ implies $1 \leqslant \mathsf{E}\,|X|^3 \leqslant (\mathsf{E}\,X^4\,\mathsf{E}\,X^2)^{1/2} = (\mathsf{E}\,X^4)^{1/2}$, whence $\rho_4 \leqslant (\mathsf{E}\,X^4)^{1/2} \leqslant \frac{\mathsf{E}\,X^4}{\mathsf{E}\,|X|^3}$ and $\rho_6 \leqslant \mathsf{E}\,|X^2 - 1|^3$; here the ρ_j's are as in (5.1.2). So, the bound in (5.1.9) is greater than that in (5.1.3) with the triple $\tau = \tilde{\tau}_{1.1}$ of constants A_3, A_4, A_6. Besides, there are a number of mistakes in the proof of (5.1.9) in [7]. It is also stated in [7], again in the i.i.d. case, that

$$|\mathsf{P}(T \leqslant z) - \Phi(z)| < \frac{36\,\mathsf{E}\,|X|^3 + 9}{\sqrt{n}}.$$

Using Stein's method, Shao [19] obtained a tighter and more general bound, also with explicit constants but without the i.i.d. assumption:

$$|\mathsf{P}(T \leqslant z) - \Phi(z)| \leqslant 10.2\gamma_2 + 25\gamma_3 \quad (5.1.10)$$

$$\leqslant 25\beta_p/\beta_2^{p/2}$$

for all $p \in [2, 3]$, with the same γ_2 and γ_3 as in (5.1.7). More recently, a Berry–Esseen bound for T was obtained in [3] for i.i.d. standard normal X_i's by means of Malliavin calculus.

Let us compare the bounds in (5.1.1) and (5.1.10). At that, let us restrict the attention to i.i.d. r.v.'s $X, X_1, \ldots, X_n$.

Consider first the case when X has a two-point zero-mean distribution, so that $\mathsf{P}(X \in \{-a, b\}) = 1$ for some positive real numbers a and b; that is,

$$\mathsf{P}(X = b) = \frac{a}{a + b} = 1 - \mathsf{P}(X = -a).$$

This case appears especially interesting, as any zero-mean distribution can be represented as a mixture of two-point zero-mean distributions—see, e.g., [13]. Without loss of generality, assume that $b \geqslant a$ and $ab = 1$. Then $b \geqslant 1$ and $\mathsf{E}\,X^2 = 1$, and hence the bound in (5.1.1) (with the triple $\tau = \tau_3$ of constants A_3, A_4, A_6) is no greater than $(16.25\rho_3 + 2.27\rho_4 + 4.83 \times 10^{-6}\rho_6)/\sqrt{n}$, where again the ρ_j's are as in (5.1.2), so that $\rho_3 = \frac{b^4 + 1}{b(b^2 + 1)}$, $\rho_4 = b - 1/b$, and $\rho_6 = (b - 1/b)^3$. On the other hand, if $b > \sqrt{n}/2$, then the bound in (5.1.10) is no less than $10.2\frac{b}{b + 1/b} \geqslant 5.1 > 1$. So, without loss of generality $b \leqslant \sqrt{n}/2$ and hence the bound in (5.1.10) equals $25\rho_3/\sqrt{n}$.

Thus (preferably with the help of the Mathematica command Reduce or similar tools), one finds that the bound in (5.1.10) will be less than the bound in (5.1.1) only if $b > 1158$; that is, only if the "asymmetry index" b/a is greater than $1158^2 = 1,340,964 > 10^6$; moreover, the inequality $b \leqslant \sqrt{n}/2$ implies that n must be no less than $(2b)^2 > (2 \times 1158)^2 = 5,363,856 > 5 \times 10^6$. One concludes that, for i.i.d. X_i's with a common two-point distribution, (5.1.1) is better than (5.1.10) unless both the sample size n and the asymmetry index are very large. Also, in the "symmetric" case when $b = a = 1$, the bound in (5.1.1) (with $\tau = \tau_3$) reduces to $1.37/\sqrt{n}$, which is $\frac{25}{1.37} > 18$ times as small as the bound in (5.1.10) (for $n \geqslant (2b)^2 = 4$).

While the two-point distributions may be of particular interest, they are of a bounded support set, and hence all their moments are finite. On the other hand, one may object that the bounds given in Theorems 5.1.1 and 5.1.2 will be infinite and hence useless if the fourth-order moments of the X_i's are infinite. However, this concern is easily addressed via truncation.

For a minute, let X denote any zero-mean r.v. If the distribution of X is continuous, then for each $b \in [0, \infty]$ there is some $a \in [0, \infty]$ such that the r.v. $X^{a,b} := X\,\mathbf{I}\{-a < X < b\}$ is zero-mean; the same holds in the case when the distribution of X is symmetric (about 0)—then one can simply take $a = b$. If the zero-mean distribution of X is not continuous or symmetric, one can use randomization, say as in [13], to still find, for each $b \in [0, \infty]$, some $a \in [0, \infty]$ and some zero-mean r.v. $X^{a,b}$ such that $\mathsf{P}(-a \leqslant X^{a,b} \leqslant b) = 1$ and $X^{a,b} = X$ on the event $\{-a < X < b\}$; let us refer to any such r.v. $X^{a,b}$ as a zero-mean truncation of the zero-mean r.v. X. (One could similarly base an appropriate construction on the so-called Winsorization $(-a) \vee (X \wedge b)$ instead of the truncation $X\,\mathbf{I}\{-a < X < b\}$.)

Now let $X_1, \ldots, X_n$ be zero-mean r.v.'s as in Theorem 5.1.1 or 5.1.2. Respectively, let $B(X_1, \ldots, X_n)$ denote (for any of the triples $\tau_1, \tau_2, \tau_3, \tilde{\tau}_1, \tilde{\tau}_2, \tilde{\tau}_3, \tilde{\tau}_{1.1}$), either one of the bounds in (5.1.1) or (5.1.3), as it depends on (the individual distributions of) the X_i's. So, $B(X_1, \ldots, X_n)$ denotes the bound in (5.1.1) under the conditions of Theorem 5.1.1, and it denotes the bound in (5.1.3) under the conditions of Theorem 5.1.2. The following corollary of Theorems 5.1.1 and 5.1.2 is immediate:

Corollary 5.1.5. *Under the conditions of Theorem 5.1.1 or 5.1.2, for each $i \in \{1, \ldots, n\}$ let $X_i^{a_i,b_i}$ be a zero-mean truncation of X_i. Then for all $z \in \mathbb{R}$*

$$|\, \mathsf{P}(T \leqslant z) - \Phi(z)| \leqslant \mathsf{P}\left(\bigcup_1^n \{X_i \notin (-a_i, b_i)\} \right) + B(X_1^{a_1, b_1}, \ldots, X_n^{a_n, b_n}).$$

$$(5.1.11)$$

Note that the upper bound in (5.1.11) can be expressed only in terms of the individual distributions of the X_i's (rather than their joint distribution), since

$$\mathsf{P}\left(\bigcup_1^n \{X_i \notin (-a_i, b_i)\} \right) = 1 - \prod_1^n \mathsf{P}\left(X_i \in (-a_i, b_i) \right).$$

So, when the bound in (5.1.1), (5.1.3), (5.1.6), or (5.1.10) can be computed, usually the "truncated" bound in (5.1.11) can be computed as well.

One may want to compare the bound in (5.1.11) with that in (5.1.10) or even with the "truncated" version of the latter bound:

$$1 - \prod_1^n \mathsf{P}\left(X_i \in (-a_i, b_i) \right) + 10.2\tilde{\gamma}_2 + 25\tilde{\gamma}_3, \qquad (5.1.12)$$

where $\tilde{\gamma}_2$ and $\tilde{\gamma}_3$ are obtained from γ_2 and γ_3 by replacing the X_i's with their zero-mean truncations $X_i^{a_i, b_i}$, as in Corollary 5.1.5.

Let us make such a comparison when the X_i's are i.i.d. with a common distribution, which is either the Student distribution with $d > 0$ degrees of freedom or the (centered) Pareto distribution with the density

$$f_s(x) := s\left(x + \frac{s}{s-1} \right)^{-s-1} \mathbf{I}\left\{ x > -\frac{1}{s-1} \right\},$$

where s is a parameter with values in the interval $(1, \infty)$. Clearly, Student's distribution with d degrees of freedom is symmetric, with heavy tails for small d and light ones for large d, whereas the Pareto distribution with parameter s is highly skewed to the right, with a heavy right tail for small $s > 1$ and a light one for large s. In keeping with the "i.i.d." assumption, let us consider the "truncated" bounds in (5.1.11) and (5.1.12) with $b_1 = \cdots = b_n =: b$ and, accordingly, $a_1 = \cdots = a_n =: a$; note that in each of the two cases under consideration (Student's or Pareto's), the value of a is uniquely determined by that of b. Then, moreover, let us (numerically) minimize the "truncated" bounds in b. The results are shown in Figs. 5.2 and 5.3. There, the graphs are shown: of the bound in (5.1.1) (longest dashes), of the minimized "truncated" bound in (5.1.11) (a bit shorter dashes), of the bound

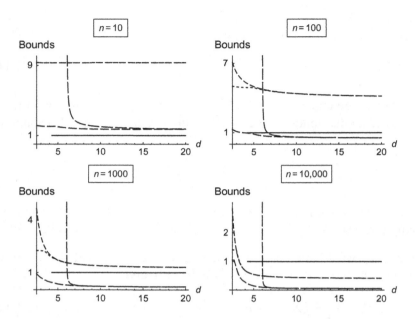

Fig. 5.2 The bounds in the case of Student's distribution with d degrees of freedom.

in (5.1.10) (yet shorter dashes), and of the minimized "truncated" bound (5.1.12) (shortest dashes)—for sample sizes $n \in \{10, 100, 1000, 10{,}000\}$, $d \in [2.5, 20]$, and $s \in [3.5, 20]$; at that, for the bounds in (5.1.1) and (5.1.11) the triple $\tau_2 = (2.02, 1.10, 0.15)$ of constant factors in (5.1.1) is used.

Figs. 5.2 and 5.3 suggest the following.

1. Predictably, the truncation helps significantly only when the tails are heavy enough—that is, for small enough values of the parameters d and s. Predictably as well, the truncation is more useful with the bound in (5.1.1) than it is with that in (5.1.10).
2. For Student's and Pareto's distributions, even the minimized "truncated" bound in (5.1.12) is nontrivial (i.e., less than 1) only if n is greater than 1000 (or even a few thousands). In fact, this bound is not much less than 0.5 even for $n = 10{,}000$ and light tails. For instance, for $n = 10{,}000$ and Student's distribution with $d = 20$ d.f., the bound in (5.1.10) and the minimized bound in (5.1.12) are both ≈ 0.417, whereas the bound in (5.1.1) and the minimized bound in (5.1.11) are both ≈ 0.055 (again, with $\tau = \tau_2 = (2.02, 1.10, 0.15)$).

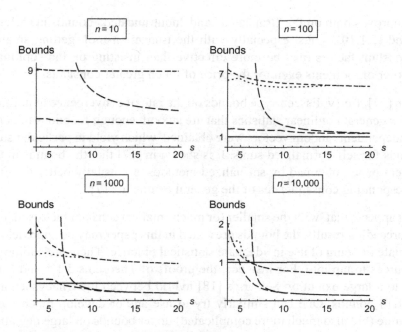

Fig. 5.3 *The bounds in the case of Pareto's distribution with parameter s.*

3. Fig. 5.3, for the Pareto case, as well as other considerations (see e.g., [12, 13] and discussion therein) suggest that the Student statistic may not be appropriate for statistical inference when the underlying distribution is significantly skewed. Alternative statistics, "correcting" for the asymmetry, were offered and considered; see [12, 13] and discussion therein.

4. If the tails are very heavy, then even the minimized "truncated," "a-bit-shorter-dashes" bound in (5.1.11) is not much less than 1 even if n is as large as 1000 and the underlying distribution is symmetric. This may be in broadly considered agreement with the fact, established in [6], that if the underlying distribution is in the domain of attraction of a stable law with index $\alpha < 2$, then the limit distribution of the self-normalized sum and, equivalently, that of the Student statistic is not normal.

5. For all considered values of n, d, and s, the minimized "truncated" bound in (5.1.11) is significantly less than that in (5.1.12). Note also that the bound in (5.1.11) can be further improved by considering triples of constants other than $\tau = \tau_2 = (2.02, 1.10, 0.15)$; for instance, one may take the minimum of the three versions of the bound in (5.1.11) corresponding to the three triples τ_1, τ_2, τ_3. Moreover, when the tails are light enough, even the "nontruncated" bound in (5.1.1) significantly

improves both on the "truncated" and "nontruncated" bounds in (5.1.12) and (5.1.10). Thus, especially with the truncation tool, getting smaller constant factors may be more effective than insisting on the "optimal" order of moments even for the price of much greater constants.

In [17], Berry–Esseen-type bounds on the rate of convergence to normality for general nonlinear statistics that are smooth enough functions of sums of independent random vectors were obtained, which yield in particular such bounds for self-normalized sums. It is shown in [17] that the bounds in the present paper, obtained by specialized methods, are usually better than the corresponding consequences of the general results in [17].

It appears that, with the smaller (or much smaller) constant factors than in the preceding results, the bounds presented in this paper may be approaching the state of being of use in adequate statistical practice. There are additional resources to tap into. For instance, the proofs of Theorems 5.1.1 and 5.1.2 rely to a large extent on Selberg's [18] hybrid between the Chebyshev and Cantelli bounds. One can similarly try to use and/or develop much more accurate (but also much more complicated) upper bounds on large deviation probabilities such as ones given and discussed in [14]; however, then the proofs can be expected to be much harder to produce or read.

5.2 PROOFS

Proof of Proposition 5.1.4. Introduce $\Lambda(a, z) := \Phi(u_{a,z})$, where $a \in (0, 1)$ and $u_{a,z} := \frac{z}{\sqrt{1+a(z^2-1)}}$, so that $\Phi_n(z) = \Lambda(\frac{1}{n}, z)$ and $\Phi(z) = \Lambda(0, z)$. By the mean value theorem, for some $b = b_z \in (0, a)$,

$$\frac{\Lambda(a, z) - \Lambda(0, z)}{a} = \frac{\partial \Lambda}{\partial a}(b, z) = \frac{u_{b,z}(1 - u_{b,z}^2)\varphi(u_{b,z})}{2(1 - b)},$$

where φ is the standard normal density function. So, to prove inequality (5.1.8), it suffices to note that $\sup_{u \in \mathbb{R}} |u(1 - u^2)\varphi(u)| = 2C$ and $\frac{1}{1-b} < \frac{1}{1-a} = \frac{n}{n-1}$ for $b \in (0, a)$ and $a = \frac{1}{n}$. That the constant factor C is the best possible in (5.1.8) follows because, by l'Hospital's rule, $\frac{\Lambda(a,z)-\Lambda(0,z)}{a} \sim \frac{\partial \Lambda}{\partial a}(a, z)$ as $a \downarrow 0$. □

The proof of Theorem 5.1.1 is based, in part, on the following two lemmas.

Lemma 5.2.1. *Take any* λ, r_*, a, b *in* $(0, \infty)$. *Take any c and r in* $(0, \infty)$ *such that*

$$c \geqslant \frac{\lambda}{r} \quad \text{and} \quad r \leqslant r_*.$$

Let Y by any r.v. such that $\mathsf{E}\,Y = 0$ *and* $\sigma := \sqrt{\mathsf{E}\,Y^2} \in (0, \infty)$. *Then*

$$\mathsf{P}(Y \geqslant c) \leqslant \psi\left(r_*, \frac{\lambda}{\sigma}\right)r, \quad \text{where} \quad \psi(u, v) := \frac{u \wedge v}{v^2 + (u \wedge v)^2}. \quad (5.2.1)$$

Also,

$$\mathsf{P}\left(Y \notin (-a, b)\right) \leqslant \frac{4\sigma^2 + (a - b)^2}{(a + b)^2}. \quad (5.2.2)$$

Proof of Lemma 5.2.1. By the condition $c \geqslant \frac{\lambda}{r}$ and Cantelli's inequality,

$$\mathsf{P}(Y \geqslant c) \leqslant \mathsf{P}(Y \geqslant \tfrac{\lambda}{r}) \leqslant \frac{\sigma^2}{\sigma^2 + (\lambda/r)^2} = \frac{r^2}{r^2 + v^2}, \quad \text{where} \quad v := \lambda/\sigma.$$

Note that $\frac{r}{r^2+v^2}$ increases in $r \in [0, v]$ and decreases in $r \in [v, \infty)$. So, if $r_* \leqslant v$, then the condition $r \leqslant r_*$ implies $\frac{r}{r^2+v^2} \leqslant \frac{r_*}{r_*^2+v^2} = \psi(r_*, v)$. If now $r_* \geqslant v$, then $\frac{r}{r^2+v^2} \leqslant \frac{v}{v^2+v^2} = \psi(r_*, v)$, so that the inequality in (5.2.1) holds in this case as well. As for inequality (5.2.2), it is due to Selberg [18]; see also, for example, [5, p. 475]. □

Lemma 5.2.2. *For any positive real numbers* x, x_1, x_2 *such that* $x \geqslant x_1 \vee x_2$, *one has*

$$x_1\overline{\Phi}(x) \leqslant \overline{\Phi}^*(x_2), \quad (5.2.3)$$

where

$$\overline{\Phi} := 1 - \Phi,$$

$$\overline{\Phi}^*(x) := 0.17\,\mathbf{I}\{0 < x < 0.752\} + x\overline{\Phi}(x)\,\mathbf{I}\{x \geqslant 0.752\}.$$

Proof of Lemma 5.2.2. It is well-known that the function $\overline{\Phi}$ is log-concave; see, for example, [4, 10]. So, the function L defined on $(0, \infty)$ by the formula $L(x) := \ln\left(x\overline{\Phi}(x)\right)$ is concave, and hence $L(x) \leqslant L(x_0) + L'(x_0)(x - x_0)$ for any x and x_0 in $(0, \infty)$. Also, $L'(0.751) > 0 > L'(0.752)$ and $L(0.752) + L'(0.752)(0.751 - 0.752) < \ln 0.17$. This implies that $L < \ln 0.17$ on $(0, \infty)$ and L is decreasing on $(0.752, \infty)$, whence $\sup\{x\overline{\Phi}(x) : x \geqslant z\} \leqslant \overline{\Phi}^*(z)$ for all $z \in (0, \infty)$. Now the lemma follows. □

Proof of Theorem 5.1.1. This proof uses some of the ideas in the proof of (5.1.9) in [7], which were previously presented in [8, 9]. As mentioned before, there are a number of mistakes of various kinds in the proof in [7]. For instance (in the notations of [7]), a bound on $\left| \Phi\left(\frac{(1-\varepsilon)r\sigma}{\sqrt{n}\,\sigma_n(r)} \right) - \Phi\left(\frac{(1-\varepsilon)r}{\sqrt{n}} \right) \right|$ analogous to that on $\left| \Phi\left(\frac{r\sigma}{\sqrt{n}\,\sigma_n(r)} \right) - \Phi\left(\frac{r}{\sqrt{n}} \right) \right|$ in [7, (1.12)] is missing there. As for the bounds in [8, 9], when the constants in them are explicit, the expressions for those bounds contain not only characteristics of the individual distributions of the X_i's but also those of their joint distribution.

Without loss of generality, assume that

$$\beta_2 = 1.$$

Take any

$$\kappa \in (0, \infty), \quad \varepsilon_4 \in (0, \tfrac{1}{2}), \quad \varepsilon_3 \in (0, \infty), \quad \varepsilon_2 \in (0, 1),$$
$$\theta_3 \in (0, 1), \quad \theta_4 \in (0, \infty) \tag{5.2.4}$$

and introduce

$$\Delta := \Delta(z) := \mathsf{P}(T \leqslant z) - \Phi(z) \tag{5.2.5}$$

and also

$$r_3 := \beta_3, \quad r_4 := \tilde{\beta}_4^{1/2}, \quad r_6 := \frac{\tilde{\beta}_6}{\beta_3^3}, \tag{5.2.6}$$

$$\varepsilon := \kappa r_4, \quad \tilde{\varepsilon}_4 := \frac{\varepsilon_4}{\kappa}. \tag{5.2.7}$$

It suffices to show that

$$|\Delta| = |\Delta(z)| \leqslant A_3 r_3 + A_4 r_4 + A_6 r_6,$$

where without loss of generality let us assume that

$$z > 0.$$

Consider the following three cases.

Case 1 ("small n"): $\varepsilon \geqslant \varepsilon_4$ or $r_3 \geqslant \varepsilon_3$.
    Note that

$$\varepsilon \geqslant \varepsilon_4 \iff r_4 \geqslant \tilde{\varepsilon}_4.$$

So,

$$|\Delta| \leqslant 1 \leqslant (A_{3,1}\, r_3) \vee (A_{4,1}\, r_4) \vee (A_{6,1}\, r_6)$$
$$\leqslant A_{3,1}\, r_3 + A_{4,1}\, r_4 + A_{6,1}\, r_6, \quad \text{where} \tag{5.2.8}$$
$$A_{3,1} := \frac{1}{\varepsilon_3}, \quad A_{4,1} := \frac{1}{\tilde{\varepsilon}_4}, \quad A_{6,1} := 0.$$

Case 2 ("large n" and "large deviations"): $\varepsilon < \varepsilon_4$, $r_3 < \varepsilon_3$, and
$z \geqslant \frac{\theta_3}{r_3} \wedge \frac{\theta_4}{r_4}$.
Then, by (5.2.1) and (5.2.5),

$$|\Delta| \leqslant (P_1 + P_2) \vee \overline{\Phi}(z),$$

where

$$P_1 := \mathsf{P}\left(T > z, V > 1 - \varepsilon_2\right)$$
$$\leqslant \mathsf{P}(S > (1 - \varepsilon_2)z) \leqslant (\psi(\varepsilon_3, \tilde{\theta}_3)r_3) \vee (\psi(\tilde{\varepsilon}_4, \tilde{\theta}_4)r_4),$$
$$\tilde{\theta}_j := (1 - \varepsilon_2)\theta_j,$$

$$P_2 := \mathsf{P}\left(V \leqslant 1 - \varepsilon_2\right) = \mathsf{P}\left(\sum_1^n (\mathsf{E}\,X_i^2 - X_i^2) \geqslant \tilde{\varepsilon}_2\right) \leqslant \psi(\tilde{\varepsilon}_4, \tilde{\varepsilon}_2)r_4,$$

$$\tilde{\varepsilon}_2 := \varepsilon_2(2 - \varepsilon_2).$$

Note also that the currently assumed case conditions $\varepsilon < \varepsilon_4$, $r_3 < \varepsilon_3$, and $z > \frac{\theta_3}{r_3} \wedge \frac{\theta_4}{r_4}$ imply $z > \frac{\theta_3}{r_3} > \frac{\theta_3}{\varepsilon_3}$ or $z > \frac{\theta_4}{r_4} > \frac{\theta_4}{\tilde{\varepsilon}_4}$. So, Lemma 5.2.2
yields

$$\overline{\Phi}(z) \leqslant \left[\overline{\Phi}^*\left(\frac{\theta_3}{\varepsilon_3}\right)\frac{r_3}{\theta_3}\right] \vee \left[\overline{\Phi}^*\left(\frac{\theta_4}{\tilde{\varepsilon}_4}\right)\frac{r_4}{\theta_4}\right].$$

Thus,

$$|\Delta| \leqslant A_{3,2}\,r_3 + A_{4,2}\,r_4 + A_{6,2}\,r_6, \quad \text{where} \tag{5.2.9}$$
$$A_{3,2} := \psi\left(\varepsilon_3, \tilde{\theta}_3\right) \vee \left[\overline{\Phi}^*\left(\frac{\theta_3}{\varepsilon_3}\right)\frac{1}{\theta_3}\right],$$
$$A_{4,2} := \left[\psi\left(\tilde{\varepsilon}_4, \tilde{\theta}_4\right) + \psi\left(\tilde{\varepsilon}_4, \tilde{\varepsilon}_2\right)\right] \vee \left[\overline{\Phi}^*\left(\frac{\theta_4}{\tilde{\varepsilon}_4}\right)\frac{1}{\theta_4}\right],$$
$$A_{6,2} := 0.$$

Case 3 ("large n" and "moderate deviations"): $\varepsilon < \varepsilon_4$, $r_3 < \varepsilon_3$, and
$z < \frac{\theta_3}{r_3} \wedge \frac{\theta_4}{r_4}$.
In this case, note that

$$\{T \leqslant z\} = \{T_z \leqslant z\}, \text{ where } T_z := S - z(\sqrt{1 + \eta} - 1) \text{ and } \eta := V^2 - 1.$$

Note also that the expression $S - z(\sqrt{1 + \eta} - 1)$ for T_z is convex in (S, η), so that its linear approximation $\left(\text{at the point } (\mathsf{E}\,S, \mathsf{E}\,\eta) = (0, 0)\right)$

$$S_z := S - z\eta/2$$

never exceeds T_z, whence

$$\delta := \frac{T_z - S_z}{z} = 1 + \frac{\eta}{2} - \sqrt{1 + \eta} \geqslant 0.$$

Therefore and because $P(T \leqslant z) = P(T_z \leqslant z)$, one has

$$P\left(S_z \leqslant (1-\varepsilon)z\right) - P(\delta > \varepsilon) \leqslant P(T \leqslant z) \leqslant P(S_z \leqslant z).$$

In view of (5.2.5), it follows that

$$\Delta \leqslant \mathrm{BE} + D(1) \quad \text{and}$$
$$-\Delta \leqslant \mathrm{BE} + P(\delta > \varepsilon) + D(1-\varepsilon) + \tilde{D}_\varepsilon,$$

where

$$\mathrm{BE} := \sup_{u \in \mathbb{R}} \left| P\left(S_z \leqslant u\right) - \Phi\left(\frac{u}{\sigma_z}\right) \right|,$$

$$\sigma_z := \sqrt{\mathsf{E}\, S_z^2} = \sqrt{\sum_1^n \mathsf{E}\, X_{i,z}^2},$$

$$D(u) := \left| \Phi(uz) - \Phi\left(\frac{uz}{\sigma_z}\right) \right|, \qquad (5.2.10)$$

$$\tilde{D}_\varepsilon := \sup_{x \geqslant 0} \left[\Phi(x) - \Phi\left((1-\varepsilon)x\right) \right]. \qquad (5.2.11)$$

Thus,

$$|\Delta| \leqslant \mathrm{BE} + P(\delta > \varepsilon) + D(1) \vee D(1-\varepsilon) + \tilde{D}_\varepsilon. \qquad (5.2.12)$$

Note also that

$$S_z = \sum_1^n X_{i,z},$$

where

$$X_{i,z} := X_i - zY_i/2 \quad \text{and} \quad Y_i := X_i^2 - \mathsf{E}\, X_i^2,$$

whence

$$\eta = \sum_1^n Y_i.$$

By a recent result of Shevtsova [20],

$$\mathrm{BE} \leqslant 0.56 \frac{\beta_{3,z}}{\sigma_z^3}, \qquad (5.2.13)$$

where

$$\beta_{3,z} := \sum_1^n E\,|X_{i,z}|^3 \leqslant \sum_1^n E\left(|X_i| + \frac{z}{2}|Y_i|\right)^3 \leqslant \frac{\beta_3}{(1-\alpha)^2} + \left(\frac{z}{2}\right)^3 \frac{\tilde{\beta}_6}{\alpha^2},$$

(5.2.14)

for any

$$\alpha \in (0,1);$$

(5.2.15)

the second inequality in (5.2.14) follows from the elementary inequality $(a+b)^3 \leqslant \frac{a^3}{(1-\alpha)^2} + \frac{b^3}{\alpha^2}$ for all a and b in $[0,\infty)$ and $\alpha \in (0,1)$. Recalling also the condition $z < \frac{\theta_3}{r_3} \wedge \frac{\theta_4}{r_4}$ and definitions (5.2.6), one has

$$\beta_{3,z} \leqslant \frac{1}{(1-\alpha)^2}\,r_3 + \frac{\theta_3^3}{8\alpha^2}\,r_6.$$

(5.2.16)

Next,

$$\sigma_z^2 = \sum_1^n E\,X_{i,z}^2 = 1 + \left(\frac{z}{2}\right)^2 \tilde{\beta}_4 - z\sum_1^n E\,X_i^3 \geqslant 1 - z\beta_3 > 1 - \theta_3. \quad (5.2.17)$$

So, (5.2.13) and (5.2.16) yield

$$\mathrm{BE} \leqslant \frac{0.56}{(1-\theta_3)^{3/2}}\left(\frac{1}{(1-\alpha)^2}\,r_3 + \frac{\theta_3^3}{8\alpha^2}\,r_6\right).$$

(5.2.18)

Further, since $0 < \varepsilon < \varepsilon_4 < \frac{1}{2}$, one has

$$\delta > \varepsilon \iff \eta \notin [2\varepsilon - 2\sqrt{2\varepsilon},\, 2\varepsilon + 2\sqrt{2\varepsilon}].$$

So, by (5.2.2), (5.2.6), and (5.2.7),

$$\mathsf{P}(\delta > \varepsilon) \leqslant \frac{4r_4^2 + 16\varepsilon^2}{32\varepsilon} = \frac{1 + 4\kappa^2}{8\kappa}\,r_4.$$

(5.2.19)

Next, by (5.2.10), for any $u \in [0,\infty)$,

$$D(u) \leqslant uz\left|\frac{1}{\sigma_z} - 1\right|\varphi\left(\frac{uz}{\sigma_z \vee 1}\right),$$

(5.2.20)

where φ is the standard normal density function. By the equalities in (5.2.17) and the case conditions $\varepsilon < \varepsilon_4$ and $z < \frac{\theta_3}{r_3} \wedge \frac{\theta_4}{r_4}$,

$$|\sigma_z^2 - 1| \leqslant z\beta_3 + \left(\frac{z}{2}\right)^2 \tilde{\beta}_4 = zr_3 + \left(\frac{z}{2}\right)^2 r_4^2 \leqslant zr_3 + \left(\frac{z}{2}\right)^2 \tilde{\varepsilon}_4 r_4 \quad (5.2.21)$$

and

$$|\sigma_z^2 - 1| \leqslant zr_3 + \left(\frac{z}{2}\right)^2 r_4^2 \leqslant \theta_3 + \theta_4^2/4.$$

(5.2.22)

Writing $\left|\frac{1}{\sigma_z} - 1\right| = \frac{|\sigma_z^2 - 1|}{\sigma_z + \sigma_z^2}$, and using (5.2.20) and (5.2.21), one has

$$D(u) \leqslant D_1(u) + D_2(u),$$

where

$$D_1(u) := r_3 \frac{v^2 \varphi(v)}{u} \rho_2, \quad D_2(u) := r_4 \frac{\tilde{\varepsilon}_4}{4} \frac{v^3 \varphi(v)}{u^2} \rho_3,$$

$$v := \frac{uz}{\sigma_z \vee 1}, \quad \rho_j := \frac{(\sigma_z \vee 1)^j}{\sigma_z + \sigma_z^2}.$$

If $\sigma_z \leqslant 1$, then by (5.2.17) for $j = 2, 3$,

$$\rho_j = \frac{1}{\sigma_z + \sigma_z^2} \leqslant \rho_* := \frac{1}{1 - \theta_3 + \sqrt{1 - \theta_3}}.$$

If $\sigma_z > 1$, then by (5.2.22) for $j = 2, 3$,

$$\rho_j = \frac{\sigma_z^j}{\sigma_z + \sigma_z^2} = \frac{1}{\sigma_z^{1-j} + \sigma_z^{2-j}} \leqslant \rho_{**,j} := \frac{1}{\sigma_*^{1-j} + \sigma_*^{2-j}},$$

where

$$\sigma_* := \sqrt{1 + \theta_3 + \theta_4^2/4}.$$

Note also that

$$\sup_{v > 0} v^j \varphi(v) = s_j := \frac{1}{\sqrt{2\pi}} \left(\frac{j}{e}\right)^{j/2}$$

for $j = 2, 3$. Therefore, recalling also the condition $\varepsilon < \varepsilon_4$, one has

$$D(1) \vee D(1 - \varepsilon) \leqslant r_3 \frac{s_2}{1 - \varepsilon_4} (\rho_* \vee \rho_{**,2}) + r_4 \frac{\tilde{\varepsilon}_4}{4} \frac{s_3}{(1 - \varepsilon_4)^2} (\rho_* \vee \rho_{**,3}).$$

$$(5.2.23)$$

Next, let us estimate $\tilde{D}_\varepsilon$. First here, one can use a special-case l'Hospital-type rule for monotonicity, such as [11, Proposition 4.1], to see that for each $x \in (0, \infty)$ the ratio $\frac{\Phi(x) - \Phi((1-t)x)}{t}$ increases in $t \in (0, 1)$. On the other hand, for each $t \in (0, 1)$ the expression $\Phi(x) - \Phi((1 - t)x)$ attains its maximum in $x \in (0, \infty)$ at $x = x_t$, where

$$x_t := \sqrt{-\frac{2 \ln(1 - t)}{t(2 - t)}}.$$

On recalling also the definition (5.2.11) of $\tilde{D}_\varepsilon$ and the conditions $0 < \varepsilon < \varepsilon_4 < \frac{1}{2}$, it follows that

$$\tilde{D}_\varepsilon \leqslant R(\varepsilon_4)\varepsilon = R(\varepsilon_4)\kappa r_4, \quad \text{where} \quad R(\varepsilon_4) := \frac{\Phi(x_{\varepsilon_4}) - \Phi((1 - \varepsilon_4)x_{\varepsilon_4})}{\varepsilon_4}.$$

$$(5.2.24)$$

Collecting (5.2.12), (5.2.18), (5.2.19), (5.2.23), and (5.2.24), one bounds $|\Delta|$ in Case 3 as follows:

$$|\Delta| \leqslant A_{3,3}\, r_3 + A_{4,3}\, r_4 + A_{6,3}\, r_6, \quad \text{where} \qquad (5.2.25)$$

$$A_{3,3} := \frac{0.56}{(1 - \theta_3)^{3/2}(1 - \alpha)^2} + \frac{s_2(\rho_* \vee \rho_{**,2})}{1 - \varepsilon_4},$$

$$A_{4,3} := \frac{1 + 4\kappa^2}{8\kappa} + \frac{\tilde{\varepsilon}_4 s_3(\rho_* \vee \rho_{**,3})}{4(1 - \varepsilon_4)^2} + R(\varepsilon_4)\kappa,$$

$$A_{6,3} := \frac{0.07}{\alpha^2}\left(\frac{\theta_3^2}{1 - \theta_3}\right)^{3/2}.$$

Collecting now the bounds (5.2.8), (5.2.9), (5.2.25) on $|\Delta|$ in Cases 1–3, one concludes that in all of the three cases

$$|\Delta| \leqslant A_3\, r_3 + A_4\, r_4 + A_6\, r_6, \quad \text{where} \qquad (5.2.26)$$
$$A_p := A_{p,1} \vee A_{p,2} \vee A_{p,3}$$

for $p = 3, 4, 6$.

Now one can arbitrarily select positive "weights" w_3, w_4, w_6 and then try numerical minimization of (say) $w_3 A_3 + w_4 A_4 + w_6 A_6$ with respect to all the parameters: α, ε_4, ε_3, ε_2, κ, θ_3, θ_4, within their specified ranges— recall (5.2.4) and (5.2.15). The target function here is complicated, and so, it is hardly possible to find the global minimum. Even though the numerical minimization is imperfect, it should be clear that the bound in (5.2.26) holds for all the allowable values of the parameters as specified in (5.2.4) and (5.2.15). The following table shows the values of the parameters α, ε_4, ε_3, ε_2, κ, θ_3, and θ_4 found by the mentioned numerical minimization for each of a few selected triples (w_3, w_4, w_6), as well as the resulting triple τ_i of the coefficients (A_3, A_4, A_6), corresponding to the so obtained values of the parameters.

Now Theorem 5.1.1 is completely proved.   □

w_3	w_4	w_6	α	ε_4	ε_3	ε_2	κ	θ_3	θ_4	Triple
1	1	1	$\frac{1013}{5000}$	$\frac{2513}{10^4}$	$\frac{387}{125}$	$\frac{2951}{10^4}$	$\frac{1187}{5000}$	$\frac{439}{1250}$	$\frac{43,307}{10^4}$	τ_1
1	1	10^6	$\frac{8067}{10^4}$	$\frac{1079}{2500}$	$\frac{34,883}{10^4}$	$\frac{1181}{10^4}$	$\frac{4739}{5000}$	$\frac{349}{10^4}$	$\frac{35,127}{5000}$	τ_2
1	10^{-4}	10^{-4}	$\frac{8}{625}$	$\frac{119}{10^4}$	$\frac{34,039}{10^4}$	$\frac{31}{5000}$	$\frac{4159}{10^4}$	$\frac{927}{2500}$	$\frac{5799}{10^4}$	τ_3

Proof of Theorem 5.1.2. This proof is quite similar to that of Theorem 5.1.1. The only essential difference that, instead of the constant 0.56 in (5.2.13) one can now use the better constant 0.4748, according to a recent result of Shevtsova [21]; then, respectively, the value $0.07 = 0.56/8$ can be replaced by the smaller value $0.4748/8$. The following table is similar to the one presented in the proof of Theorem 5.1.1.

w_3	w_4	w_6	α	ε_4	ε_3	ε_2	κ	θ_3	θ_4	Triple
1	1	1	$\frac{138}{625}$	$\frac{3177}{10^4}$	$\frac{1323}{250}$	$\frac{31}{100}$	$\frac{3357}{10^4}$	$\frac{3771}{10^4}$	$\frac{21,851}{5000}$	$\tilde{\tau}_1$
1	1	10^6	$\frac{8177}{10^4}$	$\frac{167}{500}$	$\frac{10,207}{5000}$	$\frac{1279}{10^4}$	$\frac{7021}{10^4}$	$\frac{369}{10^4}$	$\frac{36,927}{5000}$	$\tilde{\tau}_2$
1	10^{-4}	10^{-4}	$\frac{61}{5000}$	$\frac{7}{400}$	$\frac{29,837}{5000}$	$\frac{1}{200}$	$\frac{609}{10^4}$	$\frac{797}{2000}$	$\frac{38,337}{10^4}$	$\tilde{\tau}_3$
1	10	1	$\frac{2873}{10^4}$	$\frac{759}{2500}$	$\frac{10,143}{10^4}$	$\frac{276}{625}$	$\frac{186}{625}$	$\frac{809}{2000}$	$\frac{1601}{500}$	$\tilde{\tau}_{1.1}$

$\square$

REFERENCES

[1] V. Bentkus, M. Bloznelis, F. Götze, A Berry–Esséen bound for Student's statistic in the non-i.i.d. case, J. Theoret. Probab. 9 (3) (1996) 765–796.

[2] V. Bentkus, F. Götze, The Berry–Esseen bound for Student's statistic, Ann. Probab. 24 (1) (1996) 491–503.

[3] S. Bourguin, C.A. Tudor, Malliavin calculus and self normalized sums, in: Séminaire de Probabilités XLV, Lecture Notes in Math., vol. 2078, Springer, Cham, 2013, pp. 323–351.

[4] R.L. Hall, M. Kanter, M.D. Perlman, Inequalities for the probability content of a rotated square and related convolutions, Ann. Probab. 8 (4) (1980) 802–813.

[5] S. Karlin, W.J. Studden, Tchebycheff systems: with applications in analysis and statistics, in: Pure and Applied Mathematics, vol. XV, Interscience Publishers, John Wiley & Sons, New York, London, Sydney, 1966.

[6] B.F. Logan, C.L. Mallows, S.O. Rice, L.A. Shepp, Limit distributions of self-normalized sums, Ann. Probab. 1 (1973) 788–809.

[7] S.V. Nagaev, The Berry–Esseen bound for self-normalized sums, Siberian Adv. Math. 12 (3) (2003) 79–125.

[8] S.Y. Novak, On self-normalized sums, Math. Methods Stat. 9 (4) (2000) 415–436.

[9] S.Y. Novak, On self-normalized sums of random variables and the Student's statistic, Theory Probab. Appl. 49 (2) (2005) 336–344.

[10] I. Pinelis, Fractional sums and integrals of r-concave tails and applications to comparison probability inequalities, in: Advances in Stochastic Inequalities (Atlanta, GA, 1997), Contemp. Math., vol. 234, Am. Math. Soc, Providence, RI, 1999, pp. 149–168.

[11] I. Pinelis, On l'Hospital-type rules for monotonicity, J. Inequal. Pure Appl. Math. 7 (2) (2006), 19 pp., Article 40 (electronic).

[12] I. Pinelis, Exact inequalities for sums of asymmetric random variables, with applications, Probab. Theory Related Fields 139 (3/4) (2007) 605–635.

[13] I. Pinelis, Optimal two-value zero-mean disintegration of zero-mean random variables, Electron. J. Probab. 14 (26) (2009) 663–727.

[14] I. Pinelis, On the Bennett–Hoeffding inequality, Ann. Inst. Henri Poincaré Probab. Stat. 50 (1) (2014) 15–27.

[15] I. Pinelis, Exact bounds on the closeness between the Student and standard normal distributions, ESAIM Probab. Stat. 19 (2015) 24–27.

[16] I. Pinelis, R. Molzon, Optimal-order bounds on the rate of convergence to normality in the multivariate delta method, Electron. J. Stat. 10 (1) (2016) 1001–1063.

[17] I. Pinelis, R. Molzon, Optimal-order bounds on the rate of convergence to normality in the multivariate delta method (preprint), 2016, http://arxiv.org/abs/0906.0177v5, a shorter version appeared in [16].

[18] H.L. Selberg, Zwei Ungleichungen zur Erganzung des Tchebycheffschen Lemmas, Skand. Aktuarietidskr. 23 (1940) 121–125.

[19] Q.M. Shao, An explicit Berry–Esseen bound for Student's t-statistic via Stein's method, in: Stein's Method and Applications, Lect. Notes Ser. Inst. Math. Sci. Natl. Univ. Singap., vol. 5, Singapore Univ. Press, Singapore, 2005, pp. 143–155.

[20] I. Shevtsova, Refinement of estimates of the rate of convergence in the Lyapunov theorem. Dokl. Akad. Nauk 435 (1) (2010) 26–28.

[21] I. Shevtsova, On the absolute constants in the Berry–Esseen type inequalities for identically distributed summands, 2011, http://arxiv.org/abs/1111.6554.

CHAPTER 6

Sharp Probability Inequalities for Random Polynomials, Generalized Sample Cross-Moments, and Studentized Processes

Victor H. de la Peña[*], Rustam Ibragimov[†]
[*]Columbia University, New York, NY, United States
[†]Imperial College Business School and Innopolis University, Innopolis, Russia

6.1 INTRODUCTION

A number of problems in nonparametric inference in statistics and econometrics involve estimating the tail probabilities of test statistics. Several studies have discussed applications of semiparametric and nonparametric bounds for the P-values of test statistics in several contexts including nonparametric t-tests, Hotelling's T^2 test, sign tests, signed-rank tests, and permutation tests against serial dependence [18–20, 23–26, 49].

The interest in estimates for the tail probabilities of commonly used test statistics is motivated in part by the fact that the exact distributions of the statistics are frequently unknown. Even if known, the exact distributions of the test statistics are usually difficult to compute and have to be obtained by relying on computationally intensive algorithms or Monte-Carlo techniques, as in the case of permutation t-tests or linear signed-rank statistics [19, 20]. Furthermore, large sample approximations (e.g., normal approximations) require special regularity assumptions on the distribution of the observations such as existence of the second or higher moments or identical distribution (see, among others, [10, 27, 29, 48, 9, Section 5.2], the discussion in [35] and the references therein).

One should also note the importance of estimates for the tail probabilities of statistics (that hold under minimal assumptions) in the context of statistical inference in models driven by innovations with heavy tailed distributions. The latter problems are closely related to the study of robustness of statistical and econometric procedures and models in economics, finance, and related fields to nonnormality and fat-tailedness assumptions (see, e.g., [26, 42, 43], and the discussion in the books by Embrechts et al. [27],

Inequalities and Extremal Problems in Probability and Statistics. http://dx.doi.org/10.1016/B978-0-12-809818-9.00006-9

McNeil et al. [46], and Ibragimov et al. [35]). As is well documented in the empirical literature, a number of key economic and financial variables and indicators, including financial returns, have fat-tailed distributions and do not satisfy assumptions required for applications of standard inference methods (see the above monographs and references therein). In studies of the latter models with infinite variances, it is often usually assumed that the error distributions belong to the domain of attraction of stable laws. The limiting distributions for test statistics in such setups are nonstandard and involve functionals of stable processes; therefore, one has to rely on computationally intensive Monte-Carlo simulations to compute the critical values of the tests. Furthermore, the convergence of the test statistics in such models is typically very slow, hence providing inadequate approximations for finite samples (see, e.g., [1, 59] and the above references, including [9, 10, 48] for the discussion of problems with applications of autocorrelation function-based analysis in the case of heavy-tailed time series with infinite variances and higher moments, nonlinear dependence, and volatility clustering, as is typical for financial returns and foreign exchange rates).

In this chapter, we discuss sharp extensions of several (of the best known) probability and moment inequalities for sums of independent symmetric random variables (r.v.'s) to the case of random polynomials, generalized sample cross-moments, and their self-normalized and Studentized analogs. We also present extensions to the case of dependent r.v.'s through the use of measures of dependence. The results are of particular importance in situations when the observations exhibit heavy tails. We also discuss applications to self-normalized LIL's [11, 14, 15] and conservative testing procedures. The results presented in the chapter are applicable in several settings in statistics, econometrics, and time series analysis, including tests for independence and problems of detecting nonlinear dependence.

Throughout the chapter we focus on two general structures. The first concerns random polynomials and generalized sample cross-moments important in detecting nonlinear dependence and tests for independence (and their self-normalized counterparts under symmetry assumptions) in independent r.v.'s and the second involves extensions to the case of statistics in dependent r.v.'s using measures of dependence. For ease of reference, the chapter is organized as follows: Section 6.2 presents a survey of probability inequalities for sums of independent r.v's. Section 6.3 contains the main inequalities obtained in the chapter with Section 6.3.1 dealing with sharp probability inequalities for random polynomials and their self-normalized and Studentized versions in independent r.v.'s and Section 6.3.2 dealing with extensions of the results in Sections 6.2 and 6.3.1 to the case of dependent r.v.'s through measures of dependence. Section 6.3.3 presents sharp moment inequalities for random polynomials and sample cross-moments and discusses their applications. Section 6.6 is devoted to proofs.

6.2 SHARP PROBABILITY INEQUALITIES FOR SUMS OF INDEPENDENT R.V.'s AND THEIR SELF-NORMALIZED ANALOGS

Let $X_1, \ldots, X_n$ be r.v.'s on a probability space $(\Omega, \Im, P)$. A question of key interest in the calculation of P-values is to accurately estimate the tail probabilities $P\left(\sum_{i=1}^{n} X_i > x\right), x \in \mathbf{R}$. There are several results approximating tail probabilities. As examples we cite the works of Bernstein, Prokhorov, Bennett, Hoeffding, and Eaton-Pinelis [5, 6, 21, 22, 24, 33, 49, 50, 52, 57]. Among others, the book by de la Peña et al. [16] and the chapters by Shao [53] and Jing et al. [37] provide inequalities for self-normalized sums and their Studentized versions and reviews on recent developments in the area.

In what follows we present a review of the inequalities that we will be citing as well as the results for which we will provide extensions in later sections.

1. *Hoeffding's inequalities [24, 33].* Let $X_1, \ldots, X_n$ be independent r.v.'s with $EX_i = 0, i = 1, \ldots, n,$ such that $|X_i| \leq d_i \in \mathbf{R}$ (a.s.), $i = 1, \ldots, n,$ and let $D^2 = \sum_{i=1}^{n} d_i^2$. Then

$$P\left(\sum_{i=1}^{n} X_i > x\right) \leq \exp\left(-\frac{x^2}{2D^2}\right), \qquad (6.2.1)$$

$x > 0$.

Let Z be the standard normal r.v., $\phi(u) = \frac{1}{\sqrt{2\pi}} e^{-\frac{u^2}{2}}, \Phi(u) = \int_{-\infty}^{u} \phi(t) dt = \frac{1}{\sqrt{2\pi}} \int_{-\infty}^{u} e^{-\frac{t^2}{2}} dt,$ and let K be the class of twice differentiable even functions $f \colon \mathbf{R} \to \mathbf{R}$ such that f'' is nonnegative and convex and $\overline{K}$ be the class of functions $f \in K$ such that $f \colon \mathbf{R}_+ \to \mathbf{R}$ is a nondecreasing function. The classes K and $\overline{K}$ are quite wide and contain, for example, the functions $f(x) = |x|^t, t \geq 3; f(x) = (|x| - u)_+^t, t \geq 3,$ $u \geq 0$ (here and in what follows, $w_+ = \max(w, 0), w \in \mathbf{R}); f(x) = e^{h|x|},$ $h > 0,$ and $f(x) = \cosh hx, h \neq 0.$

2. *Eaton-Pinelis inequalities [21, 22, 49].* Let $X_1, \ldots, X_n$ be independent r.v.'s with $EX_i = 0, i = 1, \ldots, n$, such that $|X_i| \le d_i \in \mathbf{R}$ (a.s.), $i = 1, \ldots, n$, and let $D^2 = \sum_{i=1}^{n} d_i^2$. Then the following inequalities hold:

$$P\left(\sum_{i=1}^{n} X_i > x\right) \le \frac{1}{2} \frac{Ef(|Z|)}{f\left(\frac{x}{D}\right)}, \qquad (6.2.2)$$

$f \in \overline{K}, x > 0,$

$$P\left(\sum_{i=1}^{n} X_i > x\right) \le \inf_{0 < u < x/D} \int_u^\infty \left((t - u)^3 \middle/ \left(\frac{x}{D} - u\right)^3\right) \phi(t)dt, \qquad (6.2.3)$$

$x > 0.$

The use of the results in [25, 34] gives that for all independent mean-zero r.v.'s $X_1, \ldots, X_n$ such that $|X_i| \le d_i \in \mathbf{R}$ (a.s.), $i = 1, \ldots, n$,

$$P\left(\sum_{i=1}^{n} X_i > x\right) \le 1 - \Phi\left(\frac{x}{D} - \frac{1.5D}{x}\right), \qquad (6.2.4)$$

$x > 0$ (see the proof of Proposition 6.3.1 in this chapter). Pinelis [49] obtained the following estimates:

$$P\left(\sum_{i=1}^{n} X_i > x\right) \le \frac{2e^3}{9}\left(1 - \Phi\left(\frac{x}{D}\right)\right) \le \frac{e^3}{9}\frac{\phi\left(\frac{x}{D}\right)D}{x}, \qquad (6.2.5)$$

$x > 0$ (the second inequality in (6.2.5) was conjectured by Eaton [22]). Pinelis [49] also proposed the following alternative to (6.2.3):

$$P\left(\sum_{i=1}^{n} X_i > x\right) \le \min\left(1/2, D^2/(2x^2), \inf_{0 < u < x/D} \int_u^\infty \right.$$

$$\left. \left((t - u)^3 \middle/ \left(\frac{x}{D} - u\right)^3\right) \phi(t)dt\right), \qquad (6.2.6)$$

Dufour and Hallin [20] noted that bounds (6.2.3) and (6.2.6) can be improved when the number of the r.v.'s is taken into account and proved an inequality from which it follows that under the above conditions,

$$P\left(\sum_{i=1}^{n} X_i > x\right) \le \min\left(1/2, D^2/(2x^2), B(x/D, n)\right), \qquad (6.2.7)$$

where

$$B(y,n) = 2^{1-n} \inf_{0 \le c < y} \sum_{m=0}^{n} C_n^m f_c[(n/4)^{-1/2}(m - (n/2))]/(y - c)^3,$$

(6.2.8)

$f_c(t) = [(|t| - c)_+]^3, C_n^m = n!/(m!(n - m)!)$. Inequalities (6.2.2)–(6.2.7), with the right-hand side expressions multiplied by 2, hold for $|\sum_{i=1}^{n} X_i|$ as well.

From the results obtained by Eaton [21, 22] it also follows that the following inequality holds:

$$Ef\left(\sum_{i=1}^{n} c_i X_i\right) \le Ef(Z)$$

(6.2.9)

for all $f \in K$, independent r.v.'s $X_1, \ldots, X_n$ with $EX_i = 0, i = 1, \ldots, n$, such that $|X_i| \le 1$ (a.s.), $i = 1, \ldots, n$, and constants $c_i \in \mathbf{R}, i = 1, \ldots, n$, such that $\sum_{i=1}^{n} c_i^2 = 1$ (see also Pinelis [49]).

Edelman [24, 25] and Pinelis [49] applied inequalities (6.2.1), (6.2.3)–(6.2.5), and (6.2.9) and methods used for their proof to obtain statistically important estimates for the tail probabilities of the t-statistic and the Hotelling T^2 statistic. Dufour and Hallin [20] performed numerical comparisons of bounds of the type (6.2.3), (6.2.4), (6.2.6), and (6.2.7) and showed that estimates (6.2.6) and (6.2.7) are substantially superior to their competitors. The authors also discussed applications of the Eaton-Pinelis-type bounds to one-sample permutation t-tests, permutation t-tests against regression and against first-order autocorrelation and to testing procedures based on linear signed-rank statistics.

It is of interest to note here a relation of probability inequalities (6.2.1) and (6.2.2)–(6.2.7) to the finding by Loretan and Phillips [45, Table 1] that, for typical test sizes, the critical values of the sample split prediction test for covariance stationarity of heavy-tailed time series are lower than in the standard case of time series with innovations having forth moment. For example, from inequality (6.2.4) it follows that if $X_1, \ldots, X_n$ are independent symmetric r.v.'s (not all degenerate), then

$$P\left(\sum_{i=1}^{n} X_i \middle/ \left(\sum_{i=1}^{n} X_i^2\right)^{1/2} > x\right) \le 1 - \Phi(x - 1.5/x),$$

$x > 0$, that implies, in particular, that

$$P\left(\left(\int_0^1 dU_{\gamma/2}^s\right)^{-1} U_{\gamma/2}^s(1) > x\right) \leq 1 - \Phi(x - 1.5/x),$$

where $U_{\gamma/2}^s(1)$ is a symmetric stable process with characteristic exponent $\gamma/2, 0 < \gamma < 4$. The latter inequality implies, for example, that the critical values z_α of the sample split tests of size $\alpha\%$ for time series with innovations having Pareto-type tail behavior with tail index $\gamma, 0 < \gamma < 4$, which involve convergence to $\left(\int_0^1 dU_{\gamma/2}^s\right)^{-1} U_{\gamma/2}^s(1)$ are dominated by the quantities $\left(q_\alpha + \sqrt{q_\alpha^2 + 6}\right)/2$, where q_α is the $(1 - \alpha)\%$-quantile of the standard normal distribution: $\Phi(q_\alpha) = 1 - \alpha$.

6.3 MAIN INEQUALITIES

6.3.1 Inequalities for Random Polynomials, Generalized Sample Cross-Moments, and Their Self-Normalized and Studentized Versions in Independent R.V.'s

Let, as before, Z be the standard normal r.v., $\phi(u) = \frac{1}{\sqrt{2\pi}}e^{-\frac{u^2}{2}}$, and let $\Phi(u) = \int_{-\infty}^u \phi(t)dt = \frac{1}{\sqrt{2\pi}}\int_{-\infty}^u e^{-\frac{t^2}{2}}dt$. Let $c_i \in \mathbf{R}, r_{ki} \in \{0, 1\}, k = 1, \ldots, i - 1, i = 1, \ldots, n$, and let $X_1, \ldots, X_n$ be a sample of independent r.v.'s. Consider the random polynomials

$$V_n = \sum_{i=1}^n c_i X_1^{r_{1i}} \ldots X_{i-1}^{r_{i-1,i}} X_i \qquad (6.3.1)$$

and their self-normalized versions

$$W_n = \sum_{i=1}^n c_i X_1^{r_{1i}} \ldots X_{i-1}^{r_{i-1,i}} X_i \bigg/ \left(\sum_{i=1}^n c_i^2 X_1^{2r_{1i}} \ldots X_{i-1}^{2r_{i-1,i}} X_i^2\right)^{1/2}.$$

$$(6.3.2)$$

The class of the above polynomials V_n includes the generalized sample cross-moments $V_n^{GC-M} = \sum_{i=1}^n c_i X_{i+h_1} X_{i+h_2} \ldots X_{i+h_m}, 0 \leq h_1 < \cdots < h_m$ (in r.v.'s $X_1, \ldots, X_{n+h_m}$). The statistics V_n^{GC-M}, in turn, include, as subclasses, the sample auto-covariances $(1/n)\sum_{i=1}^n X_i X_{i-1}$ and the sample cross-moments $(1/n)\sum_{i=1}^n X_{i+h_1} X_{i+h_2} \ldots X_{i+h_m}, 0 \leq h_1 < \cdots < h_m$, arising in a number of important settings in statistics, econometrics, and time series analysis, including tests of independence and problems of detecting nonlinear dependence (see [32, 8, Section 12.1.2] for applications of sample cross-moments of order $m = 3$ and their self-normalized analogs in tests for nonlinear dependence).

The following propositions give sharp generalizations of Hoeffding- and Eaton-Pinelis-Dufour-Hallin estimates (6.2.1) and (6.2.2)–(6.2.7) to the case of random polynomials V_n and their self-normalized analogs W_n. The inequalities for the self-normalized statistics W_n hold under the only assumption of symmetry of the innovations. For example, no assumptions on boundedness of the r.v.'s or finiteness of their moments are needed. This property is central in the case of observations coming from a heavy-tailed population. At the same time, using standard symmetrization inequalities for martingales, the bounds presented in this section can also be extended to the nonsymmetric case, at the cost of increasing constants on their right-hand sides.

Proposition 6.3.1. *Let $X_1, \ldots, X_n$ be independent r.v.'s such that $EX_i = 0, |X_i| \le d_i \in \mathbf{R}$ (a.s.), $i = 1, \ldots, n$, and let $D^2 = \sum_{i=1}^{n} c_i^2 d_1^{2r_{1i}} \ldots d_{i-1}^{2r_{i-1,i}} d_i^2$, then the following inequalities hold for the random polynomials V_n defined in (6.3.1):*

$$P(V_n > x) \le \exp\left(-\frac{x^2}{2D^2}\right), \qquad (6.3.3)$$

$x > 0$,

$$P(V_n > x) \le \frac{1}{2}\frac{Ef(|Z|)}{f\left(\frac{x}{D}\right)}, \qquad (6.3.4)$$

$f \in \overline{K}, x > 0$,

$$P(V_n > x) \le \frac{e^3}{9}\frac{\phi\left(\frac{x}{D}\right)D}{x}, \qquad (6.3.5)$$

$x > \sqrt{2}D$,

$$P(V_n > x) \le \frac{2e^3}{9}\left(1 - \Phi\left(\frac{x}{D}\right)\right), \qquad (6.3.6)$$

$$P(V_n > x) \le \min(1/2, D^2/(2x^2), B(x/D, n))$$

$$\le \min\left(1/2, D^2/(2x^2), \inf_{0<u<x/D}\int_u^{\infty}\right.$$

$$\left.\left((t-u)^3 \Big/ \left(\frac{x}{D} - u\right)^3\right)\phi(t)dt\right)$$

$$\le 1 - \Phi\left(\frac{x}{D} - \frac{1.5D}{x}\right) \qquad (6.3.7)$$

$x > 0$, where $B(y, n)$ is defined in (6.2.8). The same inequalities, with the right-hand side expressions multiplied by 2, hold for $|V_n|$.

Proposition 6.3.2. Let $X_1, \ldots, X_n$ be independent symmetric r.v.'s (not all degenerate), then the following inequalities hold for the self-normalized random polynomials W_n defined in (6.3.2):

$$P(W_n > x) \leq \exp\left(-\frac{x^2}{2}\right), \tag{6.3.8}$$

$x > 0$,

$$P(W_n > x) \leq \frac{1}{2}\frac{Ef(|Z|)}{f(x)}, \tag{6.3.9}$$

$f \in \overline{K}, x > 0$,

$$P(W_n > x) \leq \frac{e^3}{9}\frac{\phi(x)}{x}, \tag{6.3.10}$$

$x > \sqrt{2}$,

$$P(W_n > x) \leq \frac{2e^3}{9}(1 - \Phi(x)), \tag{6.3.11}$$

$$P(W_n > x) \leq \min(1/2, 1/(2x^2), B(x, n))$$

$$\leq \min\left(1/2, 1/(2x^2), \inf_{0 < u < x} \int_u^\infty ((t - u)^3/(x - u)^3)\phi(t)dt\right)$$

$$\leq 1 - \Phi\left(x - \frac{1.5}{x}\right) \tag{6.3.12}$$

$x > 0$, where $B(y, n)$ is defined in (6.2.8). The same inequalities, with the right-hand side expressions multiplied by 2, hold for $|W_n|$.

The following proposition gives analogs of estimates (6.2.1) and (6.2.7) for the tail probabilities of t-statistics in the random polynomials V_n (Studentized random polynomials) that can be applied in testing serial independence of observations (see Section 6.3). Similar analogs of other exponential inequalities for sums of independent r.v.'s hold as well. The results refine and generalize those obtained by Edelman [24, 25] and Pinelis [49]. As in the case of the self-normalized random polynomials W_n, the estimates for the Studentized polynomials hold under the minimal assumption of symmetry of the underlying r.v.'s.

Proposition 6.3.3. Let $X_1, \ldots, X_n$ be independent symmetric r.v.'s (not all degenerate), V_n be as in (6.3.1) and let $\overline{V}_n = (1/n)V_n$, $s^2 = \sum_{i=1}^n (c_i X_{i+h_1} X_{i+h_2} \ldots X_{i+h_m} - \overline{V}_n)^2/(n - 1)$, then

$$P(\sqrt{n}\,\overline{V}_n/s_n > x) \le \exp(-nx^2/[2(n-1+x^2)]), \tag{6.3.13}$$

$$P(\sqrt{n}\,\overline{V}_n/s_n > x) \le \min\left(1/2, \frac{1+(x^2-1)/n}{2x^2}, B\left(\frac{x}{(1+(x^2-1)/n)^{1/2}}, n\right)\right), \tag{6.3.14}$$

$x > 0$, where $B(y,n)$ is defined in (6.2.8). The same inequalities, with the right-hand side expressions multiplied by 2, hold for $|\sqrt{n}\,\overline{V}_n/s_n|$.

6.3.2 Extensions of the Results in Sections 6.2 and 6.3.1 to the Case of Dependent R.V.'s Through Measures of Dependence

de la Peña et al. [12] obtained sharp estimates for tail probabilities and expected values of statistics in dependent r.v.'s in terms of measures of dependence of the r.v.'s.

Let $\phi_{Y_1,\ldots,Y_n}^2$ and $\delta_{Y_1,\ldots,Y_n}$ denote the following measures of dependence for absolutely continuous or discrete r.v.'s $Y_1,\ldots,Y_n$ with the one-dimensional distribution functions $F_k(y_k), k = 1,\ldots,n$, and the joint distribution function $F(y_1,\ldots,y_n)$:

$$\begin{aligned}
\phi_{Y_1,\ldots,Y_n}^2 &= \int_{-\infty}^{\infty} \cdots \int_{-\infty}^{\infty} \frac{(dF(y_1,\ldots,y_n))^2}{dF_1(y_1)\ldots dF_n(y_n)} - 1 \\
&= \int_{-\infty}^{\infty} \cdots \int_{-\infty}^{\infty} \left(\frac{dF(y_1,\ldots,y_n)}{dF_1(y_1)\ldots dF_n(y_n)}\right)^2 dF_1(y_1)\ldots dF_n(y_n) - 1
\end{aligned}$$

(multivariate analog of Pearson's ϕ^2 coefficient),

$$\delta_{Y_1,\ldots,Y_n} = \int_{-\infty}^{\infty} \cdots \int_{-\infty}^{\infty} \log\left(\frac{dF(Y_1,\ldots,Y_n)}{dF_1(y_1)\ldots dF_n(y_n)}\right) dF(y_1,\ldots,y_n)$$

(relative entropy), where the integrals are in the sense of Lebesque-Stiltjes and $\frac{dF(y_1,\ldots,y_n)}{dF_1(y_1)\ldots dF_n(y_n)}$ is to be taken to be 0 if $dF_1(y_1)\ldots dF_n(y_n) = 0$ in the former case and to be 1 if $dF_1(y_1)\ldots dF_n(y_n) = 0$ in the latter case. In the case of absolutely continuous r.v.'s $Y_1,\ldots,Y_n$ the multivariate measures $\delta_{Y_1,\ldots,Y_n}$ and $\phi_{Y_1,\ldots,Y_n}^2$ were introduced by Joe [38, 39]. In the bivariate case, the measures ϕ_{Y_1,Y_2}^2 and δ_{Y_1,Y_2} are commonly known as Pearson's ϕ^2 coefficient and the mutual information between Y_1 and Y_2, respectively. The reader is referred to, among others, [12, 36, 38–41] for the review of properties of (multivariate) dependence measures ϕ^2 and δ, related dependence concepts and inference procedures for them. If $(Y_1, \ldots, Y_n)' \sim N(\mu, \Sigma)$, then [39] $\phi^2_{Y_1, \ldots, Y_n} = |R(2I_n - R)|^{-1/2} - 1$, where I_n is the $n \times n$ identity matrix, provided that the correlation matrix R corresponding to Σ has maximum eigenvalue less than 2 and is infinite otherwise ($|A|$ denotes the determinant of a matrix A). In addition to that, if in the above case $diag(\Sigma) = (\sigma_1^2, \ldots, \sigma_n^2)$, then $\delta_{Y_1, \ldots, Y_n} = -0.5 \log(|\Sigma| / \prod_{i=1}^{n} \sigma_i^2)$. In the case of two normal r.v.'s Y_1 and Y_2 with correlation coefficient ρ, $(\phi^2_{Y_1, Y_2} / (1 + \phi^2_{Y_1, Y_2}))^{1/2} = (1 - \exp(-2\delta_{Y_1, Y_2}))^{1/2} = |\rho|$.

de la Peña et al. [12] showed that the following complete decoupling estimates hold for the tail probabilities of arbitrary statistics $h(Y_1, \ldots, Y_n)$ in r.v.'s $Y_1, \ldots, Y_n$:

$$P(h(Y_1, \ldots, Y_n) > x) \le P(h(\xi_1, \ldots, \xi_n) > x) + \phi_{Y_1, \ldots, Y_n}$$
$$(P(h(\xi_1, \ldots, \xi_n) > x))^{1/2}, \qquad (6.3.15)$$

$$P(h(Y_1, \ldots, Y_n) > x) \le \left(1 + \phi^2_{Y_1, \ldots, Y_n}\right)^{1/2} (P(h(\xi_1, \ldots, \xi_n) > x))^{1/2}, \qquad (6.3.16)$$

$$P(h(Y_1, \ldots, Y_n) > x) \le (e - 1)P(h(\xi_1, \ldots, \xi_n) > x) + \delta_{Y_1, \ldots, Y_n}, \qquad (6.3.17)$$

where $\xi_1, \ldots, \xi_n$ denote independent copies of the dependent r.v.'s $Y_1, \ldots, Y_n$. The latter results and inequalities (6.2.1) and (6.2.2)–(6.2.7) for sums of independent r.v.'s imply corresponding sharp probability inequalities for sums of dependent r.v.'s. The following sharp analogs of Hoeffding's inequality (6.2.1) and Pinelis-Dufour-Hallin estimate (6.2.7) for dependent r.v.'s hold: If $Y_1, \ldots, Y_n$ are r.v.'s with $EY_i = 0, i = 1, \ldots, n$, such that $|Y_i| \le d_i \in \mathbf{R}$ (a.s.), $i = 1, \ldots, n$, then

$$P\left(\sum_{i=1}^{n} Y_i > x\right) \le \exp\left(-\frac{x^2}{2D^2}\right) + \phi_{Y_1, \ldots, Y_n} \exp\left(-\frac{x^2}{4D^2}\right), \qquad (6.3.18)$$

$$P\left(\sum_{i=1}^{n} Y_i > x\right) \le (1 + \phi^2_{Y_1, \ldots, Y_n})^{1/2} \exp\left(-\frac{x^2}{4D^2}\right), \qquad (6.3.19)$$

$$P\left(\sum_{i=1}^{n} Y_i > x\right) \le (e - 1) \exp\left(-\frac{x^2}{2D^2}\right) + \delta_{Y_1, \ldots, Y_n}, \qquad (6.3.20)$$

$$P\left(\sum_{i=1}^{n} Y_i > x\right) \leq \min\left(1/2, D^2/(2x^2), B(x/D, n)\right)$$
$$+ \phi_{Y_1,\ldots,Y_n} \min\left(1/\sqrt{2}, D/(\sqrt{2}x), (B(x/D, n))^{1/2}\right), \qquad (6.3.21)$$

$$P\left(\sum_{i=1}^{n} Y_i > x\right) \leq (1 + \phi_{Y_1,\ldots,Y_n}^2) \min\left(1/\sqrt{2}, D/(\sqrt{2}x), (B(x/D, n))^{1/2}\right), \quad (6.3.22)$$

$$P\left(\sum_{i=1}^{n} Y_i > x\right) \leq (e-1) \min\left(1/2, D^2/(2x^2), B(x/D, n)\right) + \delta_{Y_1,\ldots,Y_n}, \qquad (6.3.23)$$

$x > 0$, where $B(y, n)$ is defined in (6.2.8).

From inequalities (6.3.15)–(6.3.17) it follows that estimates similar to those in Propositions 6.3.1–6.3.3 and involving the measures of dependence ϕ^2 and δ hold also for the random polynomials V_n and their self-normalized and Studentized analogs W_n and $\sqrt{n}\, \overline{V}_n/s_n$ in dependent r.v.'s. For example, the following propositions give analogs of inequality (6.2.7) for V_n, W_n and $\sqrt{n}\, \overline{V}_n/s_n$.

In the inequalities throughout the rest of the chapter, the extremal cases of the estimates $+\infty \leq +\infty$, $-\infty \leq +\infty$, and $-\infty \leq -\infty$ are considered to be valid inequalities; we, therefore, usually do not include assumptions on finiteness of moments of the summand r.v.'s that ensure finiteness of moments of sums of the r.v.'s into formulations of the results.

Proposition 6.3.4. *Let $X_1, \ldots, X_n$ be absolutely continuous or discrete mean-zero dependent r.v.'s such that $|X_i| \leq d_i \in R, i = 1, \ldots, n$, then the following inequalities hold for the random polynomials V_n defined in (6.3.1) in r.v.'s $X_1, \ldots, X_n$:*

$$P(V_n > x) \leq \min(1/2, D^2/(2x^2), B(x/D, n))$$
$$+ \phi_{X_1,\ldots,X_n} \min(1/\sqrt{2}, D/(\sqrt{2}x), (B(x/D, n))^{1/2}),$$
$$(6.3.24)$$

$$P(V_n > x) \leq (1 + \phi_{X_1,\ldots,X_n}^2)^{1/2} \min(1/\sqrt{2}, D/(\sqrt{2}x), (B(x/D, n))^{1/2}),$$

$$P(V_n > x) \leq (e-1) \min(1/2, D^2/(2x^2), B(x/D, n)) + \delta_{X_1,\ldots,X_n},$$

$x > 0$, *where* $D^2 = \sum_{i=1}^{n} c_i^2 d_1^{2r_{1i}} \ldots d_{i-1}^{2r_{i-1,i}} d_i^2$.

Proposition 6.3.5. *Let $X_1, \ldots, X_n$ be absolutely continuous or discrete symmetric dependent r.v.'s (not all degenerate), then the following inequalities hold for the self-normalized random polynomials W_n defined in (6.3.2) in r.v.'s $X_1, \ldots, X_n$:*

$$P(W_n > x) \leq \min(1/2, 1/(2x^2), B(x, n))$$
$$+ \phi_{X_1, \ldots, X_n} \min(1/\sqrt{2}, 1/(\sqrt{2}x), (B(x, n))^{1/2}), \quad (6.3.25)$$
$$P(W_n > x) \leq (1 + \phi^2_{X_1, \ldots, X_n})^{1/2} \min(1/\sqrt{2}, 1/(\sqrt{2}x), (B(x, n))^{1/2}),$$
$$P(W_n > x) \leq (e - 1) \min(1/2, 1/(2x^2), B(x, n)) + \delta_{X_1, \ldots, X_n},$$

$$x > 0.$$

Proposition 6.3.6. *Let $X_1, \ldots, X_n$ be symmetric r.v.'s (not all degenerate) and let*

$$\overline{V}_n = (1/n)V_n, \quad s^2 = \sum_{i=1}^{n} (c_i X_{i+h_1} X_{i+h_2} \ldots X_{i+h_m} - \overline{V}_n)^2/(n-1).$$

It follows that

$$P(\sqrt{n}\,\overline{V}_n/s_n > x) \leq \min\{1/2, (n-1+x^2)/(2nx^2), B(n^{1/2}x/(n-1+x^2)^{1/2}, n)\}$$
$$+ \phi_{X_1, \ldots, X_n} \min\{1/\sqrt{2}, (n-1+x^2)^{1/2}/((2n)^{1/2}x),$$
$$(B(n^{1/2}x/(n-1+x^2)^{1/2}, n))^{1/2}\}, \quad (6.3.26)$$
$$P(\sqrt{n}\,\overline{V}_n/s_n > x) \leq (1 + \phi^2_{X_1, \ldots, X_n})^{1/2} \min\{1/\sqrt{2}, (n-1+x^2)^{1/2}/((2n)^{1/2}x),$$
$$(B(n^{1/2}x/(n-1+x^2)^{1/2}, n))^{1/2}\},$$
$$P(\sqrt{n}\,\overline{V}_n/s_n > x) \leq (e-1) \min\{1/2, (n-1+x^2)/(2nx^2),$$
$$B(n^{1/2}x/(n-1+x^2)^{1/2}, n)\} + \delta_{X_1, \ldots, X_n},$$

$x > 0$. *The same inequalities, with the right-hand side expressions multiplied by 2, hold for $|\sqrt{n}\,\overline{V}_n/s_n|$.*

Note again that essentially only the condition on symmetry of the r.v.'s $X_1, \ldots, X_n$ is required for the estimates in Propositions 6.3.5 and 6.3.6 to hold. It is also emphasized here that bounds (6.3.24)–(6.3.26) for the statistics V_n, W_n and $\sqrt{n}\, V_n/s_n$ in dependent r.v.'s become exactly the analogs of the Eaton-Pinelis-Duffour-Hallin inequalities given by (6.3.7),

(6.3.12), and (6.3.14) in the case $\phi^2_{X_1,\ldots,X_n} = 0$, that is, in the case of the statistics in independent r.v.'s.

6.3.3 Sharp Moment Inequalities for Random Polynomials, Sample Cross-Moments, and Their Applications

The following result extends inequality (6.2.9) to the case of the random polynomials V_n in independent mean-zero r.v.'s. The results give sharp generalizations of the Khintchine-Marcinkiewicz-Zygmund-type inequalities obtained, in the case of sums of independent r.v.'s, by Eaton [21, 22], Pinelis [49], and Figiel et al. [28] (see, among others, [13, 51] and references therein for results and a review of moment inequalities for sums for independent r.v.'s and their extensions to the dependent case and [31] for their econometric applications).

Again, let K be the class of twice differentiable even functions $f : \mathbf{R} \to \mathbf{R}$ such that f'' is nonnegative and convex. In what follows, $\epsilon, \epsilon_t, t \in \{\ldots, -2, -1, 0, 1, 2, \ldots\}$, denote independent symmetric Bernoulli r.v.'s.

Proposition 6.3.7. *Let $f \in K, X_1, \ldots, X_n$ be independent r.v.'s such that $EX_i = 0, |X_i| \leq 1, i = 1, \ldots, n$, and let $\tilde{X}_1, \ldots, \tilde{X}_n$ be independent symmetric r.v.'s such that $E\tilde{X}_i^2 = 1, i = 1, \ldots, n$, then*

$$Ef(V_n) \leq Ef\left(\sum_{i=1}^{n} c_i \epsilon_i\right) \leq Ef\left(\sum_{i=1}^{n} c_i \tilde{X}_i\right). \qquad (6.3.27)$$

If, in addition to the above, $\sum_{i=1}^{n} c_i^2 = 1$, then

$$Ef(V_n) \leq Ef(Z). \qquad (6.3.28)$$

According to de la Peña et al. [13] showed, the best constants in the Khintchine-Marcinkiewicz-Zygmund inequalities for powers of generalized cross-moments V_n^{GC-M} in symmetric r.v.'s are the same as in the case of sums of independent r.v.'s. According to the following proposition, the same result holds for the Khintchine-Marcinkiewicz-Zygmund inequalities as well as for Dharmadhikari-Jogdeo-type [17] inequalities for the random polynomials V_n.

Proposition 6.3.8. *The best constants* $A_1^*(t, m), B_1^*(t, m), A_2^*(t),$ *and* $B_2^*(t)$ *in the following Khintchine-Marcinkiewicz-Zygmund inequalities*

$$A(t,m)E\left(\sum_{i=1}^{n} c_i^2 X_1^{2r_{1i}} \ldots X_{i-1}^{2r_{i-1,i}} X_i^2\right)^{t/2} \leq E\left|\sum_{i=1}^{n} c_i X_1^{r_{1i}} \ldots X_{i-1}^{r_{i-1,i}} X_i\right|^t$$

$$\leq B(t,m)E\left(\sum_{i=1}^{n} c_i^2 X_1^{2r_{1i}} \ldots X_{i-1}^{2r_{i-1,i}} X_i^2\right)^{t/2}$$

(6.3.29)

for all independent symmetric r.v.'s $X_1, \ldots, X_n$ *with finite* t-*th moment,* $t > 0$, *are given by* $A^*(t, m) = 2^{t/2-1}, 0 < t \leq t_0, A^*(t, m) = E|Z|^t, t_0 \leq t \leq 2, A^*(t, m) = 1, t \geq 2, B^*(t, m) = 1, 0 < t \leq 2, B^*(t, m) = E|Z|^t, t \geq 2$, *where* t_0 *is the nontrivial solution of the equation* $\Gamma((t_0 + 1)/2) = \Gamma(3/2), \Gamma(x)$ *is the Gamma function:* $\Gamma(x) = \int_0^{+\infty} t^{x-1}e^{-t}dt$. *The best constant* $C^*(t, m)$ *in the following Dharmadhikari-Jogdeo-type inequality*

$$E\left|\sum_{i=1}^{n} c_i X_1^{r_{1i}} \ldots X_{i-1}^{r_{i-1,i}} X_i\right|^t \leq C(t,m)n^{t/2-1}\sum_{i=1}^{n} |c_i|^t E|X_1|^{tr_{1i}} \ldots E|X_{i-1}|^{tr_{i-1,i}} E|X_i|^t$$

(6.3.30)

for all independent symmetric r.v.'s $X_1, \ldots, X_n$ *with finite* t-*th moment,* $t \geq 2$, *is given by* $C^*(t, m) = E|Z|^t$.

Using estimate (6.3.30) and Hölder's inequality, we obtain the following result that generalizes the results obtained in [3, 4] and gives a sharp estimate for the greatest order, in n, that moments of generalized moving averages $\sum_{i=1}^{n} X_{i+h_1} \ldots X_{i+h_m}$ and sample cross-moments $(1/n)\sum_{i=1}^{n} X_{i+h_1} \ldots X_{i+h_m}$ in independent mean-zero r.v.'s X_i can attain. As usual, the notation $a_n = O(b_n)$ for two nonnegative sequences (a_n) and $(b_n), n \geq 1$, means that $a_n \leq Cb_n, n \geq 1$, for some constant C that does not depend on n.

If $t_1, \ldots, t_k > 2, t = \sum_{s=1}^{k} t_s, 0 \leq h_1^{(s)} < \cdots < h_m^{(s)}, s = 1, \ldots, k; X_1, \ldots, X_{n+h_m}$, where $h_m = \max_{s=1,k} h_m^{(s)}$, are independent identically distributed r.v.'s with $EX_1 = 0$ and $E|X_1|^t < \infty$, then

$$E\prod_{s=1}^{k}\left|\sum_{i=1}^{n} X_{i+h_1^{(s)}} X_{i+h_2^{(s)}} \ldots X_{i+h_m^{(s)}}\right|^{t_s} = O(n^{t/2}).$$

The following proposition gives an estimate for the rate of convergence in the central limit theorem for moments of random polynomials $V_n = X_1^{r_{1i}} \ldots X_{i-1}^{r_{i-1,i}} X_i$ (introduced in Section 6.3.1) with equal coefficients that generalizes the classical results of [47, 58] for sums of independent r.v.'s.

Proposition 6.3.9. *If* $3 \le t < 4$, $X_1, \ldots, X_n$ *are independent identically distributed symmetric r.v.'s with* $EX_1^2 = 1$, $E|X_1|^t < \infty$, *then*

$$|E|n^{-1/2}V_n|^t - E|Z|^t| = O(n^{1-t/2}). \qquad (6.3.31)$$

6.4 APPLICATIONS IN HYPOTHESIS TESTING

In the present and the next sections, we deal with applications of the estimates considered above to several problems in statistical inference. The applications are motivated by the fact that, as discussed before, the class of random polynomials V_n includes the generalized moving averages and sample cross-moments frequently arising in statistical and econometric studies.

6.4.1 Permutation Tests Against Serial Correlation and Tests for Independence

Let $0 \le h_1 < \cdots < h_m$. Consider the problems of testing that cross-moments of a stationary time series $X_t, t = 0, 1, 2, \ldots$, with symmetric univariate distributions equal zero: H_0: $EX_{h_1}X_{h_2}\ldots X_{h_m} = 0$, for example, that the r.v.'s are uncorrelated $EX_1X_2 = 0$. These problems, arise, in particular, in tests of independence and in problems of detecting nonlinear dependence (see, for instance, [32] and Section 12.1.2 in [8] for tests for nonlinear dependence based on third moments of the time series X_t with $m = 3$). The above problems are also naturally connected to testing H_0: $\rho = 0$ against H_A: $\rho > 0$ in the first-order autoregressive model

$$X_t = \rho X_{t-1} + u_t, \qquad (6.4.1)$$

$t = 0, 1, \ldots, n$, where $u_0, u_1, \ldots, u_n$ are independent random disturbances with possibly nonidentical distributions symmetric about 0 (one evidently has $EX_{h_1}X_{h_2}\ldots X_{h_m} = 0, 0 \le h_1 < \cdots < h_m$, e.g., $EX_1X_2 = 0$ under H_0). Testing H_0 can be essentially reduced to testing that the mean of the series $Y_t = X_{t+h_1}X_{t+h_2}\ldots X_{t+h_m}, t = 1, 2, \ldots$, (respectively, $Y_t = X_t X_{t+1}, t = 1, 2, \ldots$) is zero. As in the above standard setup, the testing procedures for these problems can be based, therefore, on the t-statistics $\sqrt{n}\, \overline{V}_{n,m}^{(1)}/s_{n,m}^{(1)}$ (the superscript (1) refers to the moving average form of the statistics), where

$$\overline{V}_{n,m}^{(1)} = (1/n) \sum_{i=1}^{n} X_{i+h_1} X_{i+h_2} \cdots X_{i+h_m}$$

and

$$(s_{n,m}^{(1)})^2 = \sum_{i=1}^{n} \left(X_{i+h_1} X_{i+h_2} \cdots X_{i+h_m} - \overline{V}_{n,m}^{(1)} \right)^2 \bigg/ (n-1).$$

Evidently, under the null hypothesis, the tail probabilities of the Studentized generalized moving averages $\overline{V}_{n,m}^{(1)}/s_{n,m}^{(1)}$, for example, the Studentized sample auto-covariances

$$\overline{V}_{n,1}^{(1)}/s_{n,1}^{(1)} = \sqrt{(n-1)/n} \sum_{i=1}^{n} X_i X_{i+1} \bigg/ \left(\sum_{i=1}^{n} \left(X_i X_{i+1} - 1/n \sum_{i=1}^{n} X_i X_{i+1} \right)^2 \right)^{1/2},$$

satisfy the inequalities in Proposition 6.3.3. This implies that, when applying the above testing procedures in the latter setup one can in fact drop the terms accounting for dependence among the summands Y_t in estimates (6.5.1)–(6.5.6) (and similar generalizations of other estimates for the tail probabilities of the t-statistics in independent r.v.'s). We then have that

$$P\left(\sqrt{n}\, \overline{V}_{n,m}^{(1)} \big/ s_{n,m}^{(1)} > x \right) \le 1 - \Phi\left(n^{1/2}x/(n-1+x^2)^{1/2} \right.$$
$$\left. -1.5(n-1+x^2)^{1/2}/(n^{1/2}x) \right), \quad (6.4.2)$$

$$P\left(\sqrt{n}\, \overline{V}_{n,m}^{(1)} \big/ s_{n,m}^{(1)} > x \right) \le \frac{2e^3}{9} \left(1 - \Phi\left(n^{1/2}x/(n-1+x^2)^{1/2} \right) \right), \quad (6.4.3)$$

$$P\left(\sqrt{n}\, \overline{V}_{n,m}^{(1)} \big/ s_{n,m}^{(1)} > x \right) \le \min\left\{ 1/2, (n-1+x^2)/(2nx^2), \right.$$
$$\left. B(n^{1/2}x/(n-1+x^2)^{1/2}, n) \right\}, \quad (6.4.4)$$

where $B(y, n)$ is defined in (6.2.8). Consequently, one can use, in particular, the following conservative critical region for the test $H_0: \rho = 0$ against $H_A: \rho > 0$ with level α: $\sqrt{n}\, \overline{V}_{n,m}^{(1)}/s_{n,m} > y_\alpha^{(i)}, i = 1, 2, 3$, where $y_\alpha^{(i)}, i = 1, 2, 3$, are such that

$$1 - \Phi\left(n^{1/2}y_\alpha^{(1)} \bigg/ \left(n-1+ \left(y_\alpha^{(1)} \right)^2 \right)^{1/2} \right.$$
$$\left. - 1.5 \left(n-1+ \left(y_\alpha^{(1)} \right)^2 \right)^{1/2} \bigg/ \left(n^{1/2}y_\alpha^{(1)} \right) \right) < \alpha, \quad (6.4.5)$$

$$\frac{2e^3}{9}\left(1 - \Phi\left(n^{1/2}y_\alpha^{(2)}\bigg/\left(n - 1 + \left(y_\alpha^{(2)}\right)^2\right)^{1/2}\right)\right) < \alpha, \quad (6.4.6)$$

$$\min\left\{1/2, \left(n - 1 + \left(y_\alpha^{(3)}\right)^2\right)\bigg/\left(2n\left(y_\alpha^{(3)}\right)^2\right),\right.$$

$$\left. B\left(n^{1/2}y_\alpha^{(3)}\bigg/\left(n - 1 + \left(y_\alpha^{(3)}\right)^2\right)^{1/2}, n\right)\right\} < \alpha, \quad (6.4.7)$$

and $B(y, n)$ is defined in (6.2.8).

One can also consider the sign-based tests of independence in the time series X_t, using the fact that, under independence, $E\, sign(X_{h_1})sign(X_{h_2})\ldots$ $sign(X_{h_m}) = 0, 0 \le h_1 < \cdots < h_m$ (in particular, $E\, sign(X_1)sign(X_2) = 0$), where $sign(X_t)$ is the sign of X_t defined by $sign(X_t) = 1$ if $X_t > 0, sign(0) = 0$ and $sign(X_t) = -1$ otherwise. The latter testing procedures can be based on the statistics $\sqrt{n}\,\overline{V}_{n,m}^{(1,sign)}/s_{n,m}^{(1,sign)}$, where

$$\overline{V}_{n,m}^{(1,sign)} = (1/n)\sum_{i=1}^{n} sign(X_{i+h_1})sign(X_{i+h_2})\ldots sign(X_{i+h_m})$$

and

$$\left(s_{n,m}^{(1,sign)}\right)^2 = \sum_{i=1}^{n}\left(sign(X_{i+h_1})sign(X_{i+h_2})\ldots sign(X_{i+h_m}) - \overline{V}_{n,m}^{(1,sign)}\right)^2\bigg/(n-1).$$

Evidently, under the null hypothesis of independence, the sign versions $\sqrt{n}\,\overline{V}_{n,m}^{(1,sign)}/s_{n,m}^{(1,sign)}$ of the statistics $\overline{V}_{n,m}^{(1)}/s_{n,m}^{(1)}$ satisfy the same inequalities as above.

Note that the conservative tests based on the above Studentized statistics $\sqrt{n}\,\overline{V}_{n,m}^{(1)}/s_{n,m}$ and their sign versions $\sqrt{n}\,\overline{V}_{n,m}^{(1,sign)}/s_{n,m}^{(1,sign)}$ are equivalent to the tests based on the self-normalized moving averages

$$W_n^{(1)} = \sum_{i=1}^{n}X_{i+h_1}X_{i+h_2}\ldots X_{i+h_m}\bigg/\left(\sum_{i=1}^{n}X_{i+h_1}^2 X_{i+h_2}^2\ldots X_{i+h_m}^2\right)^{1/2}$$

and their sign versions

$$W_n^{(1,sign)} = \sum_{i=1}^{n} sign(X_{i+h_1})sign(X_{i+h_2})\ldots sign(X_{i+h_m})\bigg/$$

$$\left(\sum_{i=1}^{n}\left(sign(X_{i+h_1})sign(X_{i+h_2})\ldots sign(X_{i+h_m})\right)^2\right)^{1/2}.$$

In the case of the r.v.'s $X_1, \ldots, X_{n+h_m}$ such that $P(X_k = 0) = 0, k = 1, \ldots, n + h_m$, the sign versions of the tests are evidently equivalent to those based on the statistics

$$n^{-1/2} \sum_{i=1}^{n} sign(X_{i+h_1}) sign(X_{i+h_2}) \ldots sign(X_{i+h_m}).$$

The statistics $W_n^{(1)}$ and $W_n^{(1,sign)}$ satisfy the inequalities in Proposition 6.3.2, which imply conservative critical regions for the statistics analogous to those above for the Studentized statistics $V_n^{(1)}$ and $V_n^{(1,sign)}$.

One should note that, in the case $m = 1$, the tests based on the Studentized sample auto-covariance $\sqrt{n}\overline{V}_{n,1}^{(1)}/s_{n,1}^{(1)}$ and estimate (6.4.4) (i.e., the tests with the critical regions determined by (6.4.7)) are essentially equivalent to the permutation tests against first-order auto-regression based on the nonuniform estimates for the first-order auto-correlation coefficient proposed by Dufour and Hallin [20]. One should also emphasize here that applications of the inequalities of type (6.4.4) (e.g., the tests based on the critical regions (6.4.7)) involve the problems of determining numerically the minima of the functions in the definition of $B(y, n)$ [20] that are usually computationally intensive. On the other hand, the applications of the uniform bounds such as (6.4.2) and (6.4.3) are less computationally intensive although they yield more conservative tests than those based on (6.4.4).

If in the model (6.4.1) the disturbances u_t are dependent, one can use conservative critical regions for the tests implied by Proposition 6.3.3 and estimates (6.3.15)–(6.3.17) for statistics in dependent r.v.'s. In the latter case the statistics $\sqrt{n}\,\overline{V}_{n,m}^{(1)}/s_{n,m}^{(1)}$ satisfy inequalities analogous to (6.5.1)–(6.5.6).

The conservative tests based on the above statistics $\sqrt{n}\,\overline{V}_{n,m}^{(1)}/s_{n,m}^{(1)}$ and their sign versions $\sqrt{n}\overline{V}_{n,m}^{(1,sign)}/s_{n,m}^{(1,sign)}$ can also be applied in the problems of testing for joint independence in a sample of r.v's $X_1, \ldots, X_n$.

6.5 APPLICATIONS IN TESTING PROCEDURES FOR DEPENDENT R.V.'s

6.5.1 Conservative Critical Regions for Nonparametric t-Tests

Let $Y_1, \ldots, Y_n$ be r.v.'s with unspecified (possibly nonidentical) one-dimensional distributions symmetric about a common median μ and consider the problem of testing H_0: $\mu = \mu_0$ against $\mu > \mu_0$. The most widely used test statistic for a problem of this type is the one-sample t-statistic

$$T_n = n^{-1/2} \sum_{i=1}^{n} (Y_i - \mu_0) \Big/ \left((n-1)^{-1} \sum_{i=1}^{n} (Y_i - \bar{Y})^2 \right)^{1/2},$$

where $\bar{Y} = (1/n) \sum_{i=1}^{n} Y_i$. It is usually assumed that the r.v.'s $Y_1, \ldots, Y_n$ are independent identically distributed normal r.v.'s, in which case T_n follows a t-distribution with $n-1$ degrees of freedom under the hypothesis H_0. However, the result can no longer be used if the distributions of the Y_i's are unknown, nonidentical or if there is dependence among Y_i's. In this context, bounds for tail probabilities of T_n that hold for all symmetric r.v.'s $Y_1, \ldots, Y_n$ for different classes of dependent r.v.'s become important.

Using estimates (6.3.15)–(6.3.17) similar to [18, 24, 25] one can easily derive estimates for $P(T_n > x), x > 0$, in the case of arbitrary absolutely continuous or discrete symmetric r.v.'s $Y_1, \ldots, Y_n$ with the dependence characteristics $\phi^2_{Y_1,\ldots,Y_n}$ or $\delta_{Y_1,\ldots,Y_n}$. For example, since

$$\sum_{i=1}^{n} \left(\sum_{j=1}^{n} (Y_j - \mu_0)^2 \right)^{-1/2} (Y_i - \mu_0) = n^{1/2} T_n / (n - 1 + T_n^2)^{1/2},$$ from

inequalities (6.3.15)–(6.3.17) and estimates (6.2.1) and (6.2.7) it follows that the following bounds for the tail probabilities of the statistic T_n in r.v.'s $Y_1, \ldots, Y_n$ hold:

$$P(T_n > x) \le \exp(-nx^2/[2(n-1+x^2)])$$
$$+ \phi_{Y_1,\ldots,Y_n} \exp(-nx^2/[4(n-1+x^2)]), \qquad (6.5.1)$$

$$P(T_n > x) \le (1 + \phi^2_{Y_1,\ldots,Y_n})^{1/2} \exp(-nx^2/[4(n-1+x^2)]), \qquad (6.5.2)$$

$$P(T_n > x) \le \exp(-nx^2/[2(n-1+x^2)]) + \delta_{Y_1,\ldots,Y_n}, \qquad (6.5.3)$$

$$P(T_n > x) \le \min\{1/2, (n-1+x^2)/(2nx^2), B(n^{1/2}x/(n-1+x^2)^{1/2}, n)\}$$
$$+ \phi_{Y_1,\ldots,Y_n} \min\{1/\sqrt{2}, (n-1+x^2)^{1/2}/((2n)^{1/2}x),$$
$$(B(n^{1/2}x/(n-1+x^2)^{1/2}, n))^{1/2}\}, \qquad (6.5.4)$$

$$P(T_n > x) \le (1 + \phi^2_{Y_1,\ldots,Y_n})^{1/2} \min\{1/\sqrt{2}, (n-1+x^2)^{1/2}/((2n)^{1/2}x),$$
$$(B(n^{1/2}x/(n-1+x^2)^{1/2}, n))^{1/2}\}, \qquad (6.5.5)$$

$$P(T_n > x) \le (e-1) \min\{1/\sqrt{2}, (n-1+x^2)^{1/2}/((2n)^{1/2}x),$$
$$(B(n^{1/2}x/(n-1+x^2)^{1/2}, n))^{1/2}\} + \delta_{Y_1,\ldots,Y_n}, \qquad (6.5.6)$$

$x > 0$. Let us note that the bounds of the type (6.5.1) and (6.5.4) become exactly the estimates for $P(T_n > x)$ implied by (6.2.1) and (6.2.5)–(6.2.7) in the independent case, that is, in the case $\phi^2_{Y_1,\ldots,Y_n} = 0$. In particular, estimate (6.5.1) becomes exactly the bound for $P(T_n > x)$ in the case of independent r.v.'s $Y_1, \ldots, Y_n$ obtained by Edelman [24].

It is interesting to note that the bounds that can be derived using the above approach can be improved in the case of identically distributed, but possibly correlated, normal r.v.'s $X_1, \ldots, X_n \sim N(\mu, \Sigma), diag(\Sigma) = (\sigma^2, \ldots, \sigma^2)$, such that the correlation matrix R corresponding to Σ has a maximum eigenvalue less than 2. Namely, using the fact that in the above case, $\phi^2_{X_1,\ldots,X_n} = |R(2I_n - R)|^{-1/2} - 1$ and $\delta_{X_1,\ldots,X_n} = -0.5\log(|\Sigma|/\sigma^{2n})$ (see Section 6.1), where I_n is the $n \times n$ identity matrix, and the statistic T_n follows t-distribution with $n-1$ degrees of freedom in the case of identically distributed independent normal r.v.'s X_i, from (6.3.15)–(6.3.17) we get that, for $x > 0$,

$$P(T_n > x) \leq P(t_{n-1} > x) + (|R(2I_n - R)|^{-1/2} - 1)^{1/2}[P(t_{n-1} > x)]^{1/2},$$
$$P(T_n > x) \leq |R(2I_n - R)|^{-1/4}[P(t_{n-1} > x)]^{1/2},$$
$$P(T_n > x) \leq (e - 1)P(t_{n-1} > x) - 0.5\log(|\Sigma|/\sigma^{2n}),$$

where t_{n-1} is a t-distributed r.v. with $n - 1$ degrees of freedom. A conservative critical region for the one-sided t-test with level α is given by $T_n > y_\alpha$, where y_α is such that

$$\min\{P(t_{n-1} > y_\alpha) + (|R(2I_n - R)|^{-1/2} - 1)^{1/2}[P(t_{n-1} > y_\alpha)]^{1/2},$$
$$|R(2I_n - R)|^{-1/4}[P(t_{n-1} > y_\alpha)]^{1/2},$$
$$(e - 1)P(t_{n-1} > y_\alpha) - 0.5\log(|\Sigma|/\sigma^{2n})\} < \alpha.$$

6.5.2 Conservative Tests of Linear Hypotheses

An approach similar to that in Section 6.5.1 can be applied in other testing procedures. Consider, for example, the linear regression model $y = X\beta + u$, where X is an $n \times k$ full rank scalar matrix, $y \in \mathbf{R}^{n \times 1}; \beta \in \mathbf{R}^{k \times 1}$ is the vector of unknown parameters, and the vector of random disturbances $u \in \mathbf{R}^{n \times 1}$ has an $N(0, \Sigma)$ distribution, where, as before, $diag(\Sigma) = (\sigma^2, \ldots, \sigma^2)$, such that the correlation matrix R corresponding to Σ has a maximum eigenvalue less than 2. Suppose, further, we want to test $H_0: c'\beta = a$ against $H_A: c'\beta > a$ for some known vector $c \in \mathbf{R}^{k \times 1}$ on the basis of the t-statistic $T_c = (c'\hat{\beta} - a)/\widehat{sd}_{c'\hat{\beta}}$, where $\widehat{sd}_{c'\hat{\beta}} = (\hat{\sigma}^2 c'(X'X)^{-1}c)^{1/2}, \hat{\sigma}^2 = (y - X\hat{\beta})'(y - X\hat{\beta})/(n - k)$. Using (6.3.15)–(6.3.17) we have

$$P(T_c > x) \leq P(t_{n-k} > x) + (|R(2I_n - R)|^{-1/2} - 1)^{1/2}[P(t_{n-k} > x)]^{1/2},$$
$$P(T_c > x) \leq |R(2I_n - R)|^{-1/4}[P(t_{n-k} > x)]^{1/2},$$
$$P(T_c > x) \leq (e - 1)P(t_{n-k} > x) - 0.5\log(|\Sigma|/\sigma^{2n}),$$

where t_{n-k} denotes an r.v. with a t-distribution with $n - k$ degrees of freedom. A conservative critical region for the one-sided t-test with level α is given by $T_c > y_\alpha$, where y_α is such that

$$\min\{P(t_{n-k} > y_\alpha) + (|R(2I_n - R)|^{-1/2} - 1)^{1/2}[P(t_{n-k} > y_\alpha)]^{1/2},$$
$$|R(2I_n - R)|^{-1/4}[P(t_{n-k} > y_\alpha)]^{1/2},$$
$$(e - 1)P(t_{n-k} > y_\alpha) - 0.5\log(|\Sigma|/\sigma^{2n})\} < \alpha.$$

Using the asymptotic properties of the statistics T_n and T_c, we conclude that the above critical regions can also be used in the case of non-Gaussian identically distributed errors when the sample size n is sufficiently large.

6.6 PROOFS

The proofs of Propositions 6.3.1–6.3.8 are based on the following reduction properties (Lemmas 6.6.2 and 6.6.3) for martingales and, more generally, multiplicative systems of an arbitrary order proved in [12, 55].

Definition 6.6.1. R.v.'s $X_1, \ldots, X_n$ form a multiplicative system of order $\alpha \in \mathbf{N}$ (shortly, $MS(\alpha)$) if $E|X_j|^\alpha < \infty, j = 1, \ldots, n$, and for any $\alpha_j \in \{0, 1, \ldots, \alpha\}, j = 1, \ldots, n$,

$$E \prod_{j=1}^{n} X_j^{\alpha_j} = \prod_{j=1}^{n} EX_j^{\alpha_j}.$$

The systems $MS(1)$ and $MS(2)$ under the names multiplicative and strongly multiplicative systems, respectively, were introduced by Alexits [2]. Multiplicative systems of an arbitrary order were considered, for example, by Kwapien [44] and Sharakhmetov [54]. Examples of the multiplicative systems of order 1 $MS(1)$ are given, besides independent r.v.'s, by the lacunary trigonometric systems $\{\cos 2\pi n_k x, \sin 2\pi n_k x, k = 1, 2, \ldots\}$ on the interval $[0, 1]$ with the Lebesque measure for $n_{k+1}/n_k \geq 2$ and also by martingale-difference sequences. Examples of the systems $MS(2)$ are given by the lacunary trigonometric systems for $n_{k+1}/n_k \geq 3$ and martingale-difference sequences $X_1, \ldots, X_n$ with the nonrandom

conditional variances $E(X_n^2|X_1,\ldots,X_{n-1}) = b_n^2 \in \mathbf{R}, n = 1, 2, \ldots$.
Examples of the systems $MS(\alpha)$ include, for instance, the lacunary
trigonometric systems with large lacunas, that is, with $n_{k+1}/n_k \geq \alpha + 1$
and also ϵ-independent and asymptotically independent r.v.'s introduced by
Zolotarev [60] (see the discussion in [12]).

Lemma 6.6.2. *Let $\alpha \in N$, and let $A_i, i = 1,\ldots,n$, be sets of real
numbers such that $card(A_i) \leq \alpha + 1, i = 1,\ldots,n$. R.v.'s $X_1,\ldots,X_n$
taking values in $A_1,\ldots,A_n$, respectively, form a multiplicative system of
order α if and only if they are jointly independent.*

Lemma 6.6.2 implies the following reduction property for general
martingale-difference sequences.

Lemma 6.6.3. *A sequence of r.v.'s $\{X_n\}$ on a probability space
$(\Omega, \Im, P)$ assuming two values is a martingale-difference with respect to
an increasing sequence of σ-algebras $\Im_0 = (\Omega, \emptyset) \subseteq \Im_1 \subseteq \ldots \subseteq \Im$ if and
only if the r.v.'s $\{X_n\}$ are jointly independent.*

Since the r.v.'s $\eta_i = \epsilon_1^{r_{1i}} \ldots \epsilon_{i-1}^{r_{i-1,i}} \epsilon_i, i = 1,\ldots,n$, form a martingale-
difference sequence with respect to the σ-algebras $\sigma(\epsilon_1, \epsilon_2, \ldots, \epsilon_i)$,
$i = 1,\ldots,n$, we get the following corollary of Lemma 6.6.3 that describes
the reduction properties of the summands in the random polynomials V_n in
independent symmetric r.v.'s.

Lemma 6.6.4. *The r.v.'s $\eta_1,\ldots,\eta_n$ are jointly independent.*

Remark 6.6.5. Let us note that Lemma 6.6.3 also implies Propositions 1 and 3 in [7]. Let $w(z) = 1$ for $z \geq 0$, and $w(z) = 0$ for
$z < 0$, and let $X_0,\ldots,X_{n-1}$ and $Y_1,\ldots,Y_n$ be r.v.'s such that Y_t is
independent of $\sigma(X_0, X_1,\ldots,X_{t-1}, Y_1,\ldots,Y_{t-1})$ for each $t = 1,\ldots,n$,
and $P(Y_t > 0) = P(Y_t < 0) = 1/2$ for $t = 1,\ldots,n$. Also, let $g_t = g_t(X_0, X_1,\ldots,X_t, Y_1,\ldots,Y_t), t = 0,\ldots,n-1$, be a sequence of
measurable functions of $X_0,\ldots,X_t$ such that $P(g_t = 0) = 0$ for
$t = 0,\ldots,n-1$. According to Propositions 1 and 3 in [7], the statistic $S_g = \sum_{t=1}^n w(Y_t g_{t-1})$ has a binomial distribution $Bi(n, 0.5)$ with parameters
n, 0.5, and, moreover, the r.v.'s $w(Y_t g_{t-1}), t = 1,\ldots,n$, are jointly
independent if the r.v.'s $Y_1,\ldots,Y_n$ have continuous symmetric distributions.
To see that the latter results follow from Lemma 6.6.3, it suffices to observe
that the r.v.'s $2w(Y_t g_{t-1}) - 1$ form a martingale-difference sequence
with respect to the σ-algebras $\sigma(X_0, X_1,\ldots,X_t, Y_1,\ldots,Y_t)$ under the

assumptions of the propositions. In a recent paper, So and Shin [56] considered sign tests for random walks against stationary alternative hypothesis in the model $y_t = h(x_t), x_t = \rho(x_{t-1}, \ldots, x_{t-k}) + u_t, t = 1, \ldots, n$, where $\{y_t\}, t = 0, \ldots, n$, is a set of observations, $h(x_t)$ is an unknown monotone transformation of $\{x_t\}, \rho(x_k, \ldots, x_1)$ is an unknown regression function of interest, k is a positive integer, and $\{u_t\}$ is a sequence of errors satisfying the conditions.

A1: $\{sign(u_t)\}$ is a martingale difference sequence with respect to an increasing sequence of σ-fields $\{\Im_t\}, t = 1, \ldots, n,$.

A2: $P(u_t = 0|\Im_{t-1}) = 0$, where $sign(u_t)$ is the sign of u_t defined by $sign(u_t) = 1$ if $u_t > 0, sign(0) = 0$ and $sign(u_t) = -1$ otherwise. From the above, it follows that in fact the conditions A1 and A2 are equivalent to joint independence of the signs $sign(u_t)$. Moreover, using Lemma 6.6.3, one also immediately gets that the r.v.'s $sign(u_t)sign(v_{t-1}), t = 1, \ldots, n$, where $v_t, t = 1, \ldots, n$, is a sequence of $\Im_t$-measurable r.v.'s with no atom at zero, are jointly independent. This implies results in [56] concerning distributional properties of the test statistic based on the quantity $S_n(\rho) = \sum_{t=1}^{n} sign(u_t(\rho))sign(v_{t-1})$, where $u_t(\rho) = x_t - \rho(x_{t-1}, \ldots, x_{t-k})$.

Proof of Propositions 6.3.1–6.3.8. Let $r_{ki} \in \{0, 1\}, k = 1, \ldots, i - 1$, $i = 1, \ldots, n$. Further, let ϵ be a symmetric Bernoulli r.v. According to [34], $Eg(Y) \leq Eg(\epsilon)$ for all continuous convex functions $g: [-1, 1] \rightarrow \mathbf{R}$ and all r.v.'s Y such that $EY = 0$ and $|Y| \leq 1$. The inequality implies that $Ef(cY + d) \leq Ef(ca\epsilon + d)$ for all continuous convex functions $f: \mathbf{R} \rightarrow \mathbf{R}$, constants $a, c, d \in \mathbf{R}$, and all r.v.'s Y such that $EY = 0$ and $|Y| \leq 1$. Using this fact, conditioning arguments and Lemma 6.6.4, we get that if $c_i \in \mathbf{R}, r_{ki} \in \{0, 1\}, k = 1, \ldots, i - 1, i = 1, \ldots, n$, and $X_1, \ldots, X_n$ are independent r.v.'s such that $EX_i = 0, |X_i| \leq 1$ (a.s.), $i = 1, \ldots, n$, then

$$Ef\left(\sum_{i=1}^{n} c_i X_1^{r_{1i}} \ldots X_{i-1}^{r_{i-1,i}} X_i\right) \leq Ef\left(\sum_{i=1}^{n} c_i \epsilon_1^{r_{1i}} \ldots \epsilon_{i-1}^{r_{i-1,i}} \epsilon_i\right) = Ef\left(\sum_{i=1}^{n} c_i \epsilon_i\right)$$

(6.6.1)

for all continuous convex functions $f: \mathbf{R} \rightarrow \mathbf{R}$. From Corollary 2.5 in [49] and Theorem 1.1 in [28] it follows that

$$Ef\left(\sum_{i=1}^{n} c_i \epsilon_i\right) \leq Ef\left(\sum_{i=1}^{n} c_i X_i'\right)$$

(6.6.2)

for all twice differentiable even functions $f\colon \mathbf{R} \to \mathbf{R}$ such that f'' is convex and all independent symmetric r.v.'s X'_i such that $EX'^2_i = 1, i = 1, \ldots, n$. Inequalities (6.6.1) and (6.6.2) imply (6.3.27). Inequality (6.3.28) follows letting the r.v.'s $\tilde{X}_i$ in (6.3.27) be the standard normal r.v.'s. Let us prove inequalities (6.3.3)–(6.3.7). Let $X_1, \ldots, X_n$ be independent r.v.'s such that $EX_i = 0, |X_i| \le d_i, i = 1, \ldots, n$, and let $D^2 = \sum_{i=1}^n c_i^2 d_1^{2r_{1i}} \ldots d_2^{2r_{i-1,i}} d_i^2$. By Chebyshev's inequality we have

$$P(V_n > x) \le \exp(-hx) E \exp(hV_n), \tag{6.6.3}$$

$x > 0, h > 0$. Using the above Hunt's inequality and Lemma 6.6.4, we get

$$E \exp(hV_n) \le E \exp\left(h \sum_{i=1}^n c_i d_1^{r_{1i}} \ldots d_{i-1}^{r_{i-1,i}} d_i \epsilon_i \right). \tag{6.6.4}$$

According to [33],

$$E \exp\left(h \sum_{i=1}^n X_i \right) \le \exp\left(\frac{1}{2} h^2 \sum_{i=1}^n d_i^2 \right) \tag{6.6.5}$$

for all independent r.v.'s $X_1, \ldots, X_n$ such that $EX_i = 0, |X_i| \le d_i \in \mathbf{R}$. From (6.6.3) to (6.6.5) it follows that for $x > 0$

$$P(V_n > x) \le \exp\left(\frac{1}{2} h^2 D^2 - hx \right). \tag{6.6.6}$$

The right-hand side of (6.6.6) has its minimum at $h = \frac{x}{D^2}$. Inserting this value in (6.6.6) we obtain inequality (6.3.3). From Chebyshev's inequality it follows that

$$P(V_n > x) = \frac{1}{2} P(|V_n| > x) \le \frac{1}{2} \frac{Ef\left(\frac{1}{D}|V_n|\right)}{f\left(\frac{x}{D}\right)} \tag{6.6.7}$$

for all $f \in \overline{K}$ (introduced in Section 6.1). By (6.3.28),

$$Ef\left(\frac{1}{D}|V_n| \right) \le Ef(|Z|), \tag{6.6.8}$$

$f \in K$. Inequalities (6.6.7) and (6.6.8) imply (6.3.4). From the results obtained by Pinelis [49] (see also Pinelis [50]) it follows that inequality (6.6.8) implies that $P(V_n > x) \le \frac{2e^3}{9}(1 - \Phi(\frac{x}{D})), x > 0; P(V_n > x) \le \frac{e^3}{9} \frac{\phi(\frac{x}{D})D}{x}, x > \sqrt{2}$, that is, (6.3.5) and (6.3.6) hold (note that the latter inequalities follow directly from Theorem 5.4 in [50] and the martingale structure of V_n). From (6.6.1), the fact that the function $f(x) = (|x| - u)^3_+$

belongs to $\overline{K}$ and the results obtained by Eaton [21, 22] (see also Dufour and Hallin [20]), it follows that

$$E\left[\left(\frac{1}{D}|V_n| - u\right)_+\right]^3 \leq E\left[\left(\frac{1}{D}\left|\sum_{i=1}^{n} c_i d_1^{r_{1i}} \ldots d_{i-1}^{r_{i-1,i}} d_i \epsilon_i\right| - u\right)_+\right]^3$$

$$\leq E\left[\left(\left|\frac{1}{\sqrt{n}}\sum_{i=1}^{n}\epsilon_i\right| - u\right)_+\right]^3. \qquad (6.6.9)$$

Moreover,

$$E\left(\frac{1}{D}V_n\right)^2 \leq \frac{1}{D^2}\sum_{i=1}^{n} c_i^2 d_1^{2r_{1i}} \ldots d_{i-1}^{2r_{i-1,i}} d_i^2 E\epsilon_i^2 = 1. \qquad (6.6.10)$$

Similarly to the proof of Proposition 1 in [20], relations (6.6.9) and (6.6.10) and Chebyshev's inequality give the first estimate in (6.3.7). The fact that $E\left[\left(\frac{1}{\sqrt{n}}|\sum_{i=1}^{n}\epsilon_i| - u\right)_+\right]^3 \leq E[(|Z| - u)_+]^3$ by (6.2.9) and the first estimate in (6.3.7) imply the second inequality in (6.3.7). Since, according to [25],

$$\inf_{0<u<x/D_j} \int_u^\infty \left((t-u)^3 \bigg/ \left(\frac{x}{D_j} - u\right)^3\right) \phi(t)dt \leq 1 - \Phi\left(x - \frac{1.5}{x}\right),$$

$$\qquad (6.6.11)$$

$x > 0$, we get the last estimate in (6.3.7).

Conditioning on $|X_1|, \ldots, |X_n|$ and using estimates (6.3.4)–(6.3.7) we obtain (similarly to [24, 25]) inequalities (6.3.8)–(6.3.12).

Let us prove Proposition 6.3.3. Using the relation

$$W_n = \sqrt{n} \frac{\overline{V}_n/s_n}{\left(1 - 1/n + \overline{V}_n^2/s_n^2\right)^{1/2}},$$

we obtain, similarly to [24, 25]

$$P\left(\sqrt{n}\,\overline{V}_n/s_n > x\right) = P\left(W_n > \frac{x}{(1 + (x^2 - 1)/n)^{1/2}}\right).$$

This and Proposition 6.3.2 imply the inequalities in Proposition 6.3.3.

The estimates in Propositions 6.3.4–6.3.6 follow from Propositions 6.3.1–6.3.3 and inequalities (6.3.15)–(6.3.17).

The expressions for the best constants in inequalities (6.3.29) follow from Lemma 6.6.4 and the results obtained by Haagerup [30]. The right-hand side inequality (6.3.29) and the estimate $\left(\sum_{i=1}^{n} z_i\right)^{t/2} \leq \sum_{i=1}^{n} n^{t/2-1} z_i^{t/2}$ for all $z_1, \ldots, z_n \geq 0, t \geq 2$, imply that estimate (6.3.30) holds with the constant $C^*(t, m)$ defined in Proposition 6.3.8. Sharpness of the constant $C^*(t, m)$ follows from the choice $c_i = 1/\sqrt{n}, i = 1, \ldots, n$, $X_i = \epsilon_i, i = 1, \ldots, n + h_m$, Lemma 6.6.4 and the central limit theorem.

Using conditioning arguments and the inequality $Ef(ca\epsilon + d) \leq Ef(cX + d)$ for all twice differentiable even functions $f \colon \mathbf{R} \to \mathbf{R}$ such that f'' is convex, constants $a, c, d \in \mathbf{R}$, and all symmetric r.v.'s X such that $EX^2 = a^2$ implied by Corollary 2.5 in [49] and Theorem 1.1 in [28], by induction and Lemma 6.6.4 we get

$$E\left|\sum_{i=1}^{n} X_1^{r_{1i}} \ldots X_{i-1}^{r_{i-1,i}} X_i\right|^t \geq E\left|\sum_{i=1}^{n} \epsilon_1^{r_{1i}} \ldots \epsilon_{i-1}^{r_{i-1,i}} \epsilon_i\right|^t = E\left|\sum_{i=1}^{n} \epsilon_i\right|^t \quad (6.6.12)$$

for independent identically distributed symmetric r.v.'s $X_1, \ldots, X_{n+h_m}$ with $EX_1^2 = 1, E|X_1|^t < \infty, t \geq 3$. Moreover, from estimate (3.28) in [13] and Lemma 6.6.4 it follows that

$$E\left|\sum_{i=1}^{n} X_1^{r_{1i}} \ldots X_{i-1}^{r_{i-1,i}} X_i\right|^t \leq E\left|\sum_{i=1}^{n} \epsilon_i\right|^t + O(n) \quad (6.6.13)$$

for independent identically distributed symmetric r.v.'s $X_1, \ldots, X_{n+h_m}$ with $EX_1^2 = 1, E|X_1|^t < \infty, 2 < t \leq 4$. Relations (6.6.12) and (6.6.13) and the fact that, according to [47, 58], $\left|E\left|\frac{1}{\sqrt{n}}\sum_{i=1}^{n} \epsilon_i\right|^t - E|Z|^t\right| = O(n^{1-t/2}), 3 \leq t < 4$, imply relation (6.3.31). $\qquad \square$

Acknowledgments

The authors are grateful to an anonymous referee; the Editor, Iosif Pinelis; and the participants at the 30th European Meeting of Statisticians (June 2015) for helpful comments on the chapter and their work on inequalities. Research of R. Ibragimov is supported by a grant from the Russian Science Foundation (Project No. 16-18-10432).

REFERENCES

[1] R.J. Adler, R.E. Feldman, C. Gallagher, Analysing stable time series, in: R.J. Adler, R.E. Feldman, M.S. Taqqu (Eds.), A Practical Guide to Heavy Tails, Birkhauser, Boston, MA, 1998, pp. 113–158.

[2] G. Alexits, Convergence problems of orthogonal series, in: International Series of Monographs in Pure and Applied Mathematics, vol. 20, Pergamon Press, New York, Oxford, Paris, 1961.

[3] O.D. Anderson, Exact general-lag serial correlation moments and approximate low-lag partial correlation moments for Gaussian White noise, J. Time Ser. Anal. 14 (1993) 551–574.

[4] O.D. Anderson, Z.G. Chen, Higher order moments of sample autocovariances and sample autocorrelations from an independent time series, J. Time Ser. Anal. 17 (1996) 323–331.

[5] G. Bennett, Probability inequalities for the sum of independent random variables, J. Am. Stat. Assoc. 57 (1962) 33–45.

[6] S. Bernstein, Theory of Probability, Gostekhizdat, Moscow, 1946 (in Russian).

[7] B. Campbell, J.M. Dufour, Exact nonparametric orthogonality and random walk tests, Rev. Econ. Stat. 77 (1995) 1–16.

[8] J.Y. Campbell, A.W. Lo, A.C. MacKinlay, The Econometrics of Financial Markets, Princeton University Press, Princeton, NJ, 1997.

[9] R. Cont, Empirical properties of asset returns: stylized facts and statistical issues, Quant. Finan. 1 (2001) 223–236.

[10] R.A. Davis, T. Mikosch, The sample autocorrelations of heavy-tailed processes with applications to ARCH, Ann. Stat. 26 (1998) 2049–2080.

[11] V.H. de la Peña, A general class of exponential inequalities for martingales and ratios, Ann. Probab. 27 (1999) 537–564.

[12] V.H. de la Peña, R. Ibragimov, S. Sharakhmetov, Characterizations of joint distributions, copulas, information, dependence and decoupling, with applications to time series, in: J. Rojo (Ed.), 2nd Erich L. Lehmann Symposium Optimality, IMS Lecture Notes Monograph Series, vol. 49, 2006, pp. 183–209, doi:10.1214/074921706000000455.

[13] V.H. de la Peña, R. Ibragimov, S. Sharakhmetov, On extremal distributions and sharp L_p-bounds for sums of multilinear forms, Ann. Probab. 31 (2003) 630–675.

[14] V.H. de la Peña, M.J. Klass, T.L. Lai, Moment bounds for self-normalized processes, in: High Dimensional Probability, II (Seattle, WA, 1999), Progress in Probability, Birkhäuser, Boston, MA, 2000, pp. 3–12.

[15] V.H. de la Peña, M.J. Klass, T.L. Lai, Self-normalized processes: exponential inequalities, moment bounds and iterated logarithm laws, Ann. Probab. 32 (2004) 1902–1933.

[16] V.H. de la Peña, T.L. Lai, Q.M. Shao, Self-Normalized Processes: Limit Theory and Statistical Applications, Springer, New York, NY, 2009.

[17] S.W. Dharmadhikari, K. Jogdeo, Bounds on moments of certain random variables, Ann. Math. Stat. 40 (1969) 1506–1509.

[18] J.M. Dufour, M. Hallin, Nonuniform bounds for nonparametric t-tests, Economet. Theor. 7 (1991) 253–263.

[19] J.M. Dufour, M. Hallin, Simple exact bounds for distributions of linear signed rank statistics, J. Stat. Plan. Inference 31 (1992) 311–333.

[20] J.M. Dufour, M. Hallin, Improved Eaton bounds for linear combinations of bounded random variables, with statistical applications, J. Am. Stat. Assoc. 88 (1993) 1026–1033.

[21] M.L. Eaton, A note on symmetric Bernoulli random variables, Ann. Math. Stat. 41 (1970) 1223–1226.

[22] M.L. Eaton, A probability inequality for linear combinations of bounded random variables, Ann. Stat. 2 (1974) 609–614.

[23] M.L. Eaton, B. Efron, Hotelling's T^2 test under symmetry conditions, J. Am. Stat. Assoc. 65 (1970) 702–711.

[24] D. Edelman, Bounds for a nonparametric t table, Biometrika 73 (1986) 242–243.

[25] D. Edelman, An inequality of optimal order for the tail probabilities of the t-statistic under symmetry, J. Am. Stat. Assoc. 85 (1990) 120–122.

[26] B. Efron, Student's t-test under symmetry conditions, J. Am. Stat. Assoc. 64 (1969) 1278–1302.

[27] P. Embrechts, C. Klüppelberg, T. Mikosch, Modelling Extremal Events for Insurance and Finance, Springer, New York, NY, 1997.

[28] T. Figiel, P. Hitczenko, W.B. Johnson, G. Schechtman, J. Zinn, Extremal properties of Rademacher functions with applications to the Khintchine and Rosenthal inequalities, Trans. Am. Math. Soc. 349 (1997) 997–1027.

[29] C.W.J. Granger, D. Orr, Infinite variance and research strategy in time series analysis, J. Am. Stat. Assoc. 67 (1972) 275–285.

[30] U. Haagerup, The best constants in the Khintchine inequality, Stud. Math. 70 (1982) 231–283.

[31] B.E. Hansen, The integrated mean squared error of series regression and a Rosenthal Hilbert-space inequality, Economet. Theor. 31 (2015) 337–361.

[32] D.A. Hsieh, Testing for nonlinear dependence in daily foreign-exchange rates, J. Bus. 62 (1989) 339–368.

[33] W. Hoeffding, Probability inequalities for sums of bounded random variables, J. Am. Stat. Assoc. 58 (1963) 13–30.

[34] G.A. Hunt, An inequality in probability theory, Proc. Am. Math. Soc. 6 (1955) 506–510.

[35] M. Ibragimov, R. Ibragimov, J. Walden, Heavy-Tailed Distributions and Robustness in Economics and Finance, Springer, New York, NY, 2015.

[36] R. Ibragimov, A. Prokhorov, Heavy Tails and Copulas: Topics in Dependence Modelling in Economics and Finance, World Scientific, Singapore, 2016 (forthcoming).

[37] B.Y. Jing, Q.M. Shao, Q. Wang, Self-normalized Cramér-type large deviations for independent random variables, Ann. Probab. 31 (2003) 2167–2215.

[38] H. Joe, Majorization, randomness and dependence for multivariate distributions, Ann. Probab. 15 (1987) 1217–1225.

[39] H. Joe, Relative entropy measures of multivariate dependence, Proc. Am. Math. Soc. 84 (1989) 157–164.

[40] H. Joe, Multivariate Models and Dependence Concepts, Chapman & Hall, London, 1997.

[41] H. Joe, Dependence Modeling With Copulas, Chapman & Hall/CRC, London, 2014.

[42] J. Jurečkovà, P.K. Sen, Robust Statistical Procedures: Asymptotics and Interrelations, Wiley Series in Probability and Statistics: Applied Probability and Statistics, A Wiley-Interscience Publication, John Wiley & Sons, New York, NY, 1996.

[43] T. Kariya, B.K. Sinha, Robustness of Statistical Tests. Statistical Modeling and Decision Science, Academic Press, Boston, MA, 1989.

[44] S. Kwapien, Decoupling inequalities for polynomial chaos, Ann. Probab. 15 (1987) 1062–1071.

[45] M. Loretan, P.C.B. Phillips, Testing the covariance stationarity of heavy-tailed time series, J. Empir. Finance 3 (1994) 211–248.

[46] A.J. McNeil, R. Frey, P. Embrechts, Quantitative Risk Management: Concepts, Techniques, and Tools, Princeton University Press, Princeton, NJ, 2005.

[47] R. Michel, Nonuniform central limit bounds with applications to probabilities of deviations, Ann. Probab. 4 (1976) 102–106.

[48] T. Mikosch, C. Stărică, Limit theory for the sample autocorrelations and extremes of a GARCH(1, 1) process, Ann. Stat. 28 (2000) 1427–1451.

[49] I. Pinelis, Extremal probabilistic problems and Hotelling's T^2 test under a symmetry condition, Ann. Stat. 22 (1994) 357–368.

[50] I. Pinelis, Optimal tail comparison based on comparison of moments, in: High Dimensional Probability (Oberwolfach, 1996), Progress in Probability, vol. 43, Birkhäuser, Basel, 1998, pp. 297–314.

[51] I. Pinelis, Exact Rosenthal-type bounds, Ann. Probab. 43 (2015) 2511–2544.

[52] Y.V. Prokhorov, An extremal problem in probability theory, Theory Probab. Appl. 4 (1959) 201–203.

[53] Q.M. Shao, Cramér-type large deviation for Students t statistic, J. Theor. Probab. 12 (1999) 387–398.

[54] S. Sharakhmetov, R-independent random variables and multiplicative systems, Dopov. Dokl. Akad. Nauk Ukraïni (1993) 43–45 (Russian).

[55] S. Sharakhmetov, R. Ibragimov, A characterization of joint distribution of two-valued random variables and its applications, J. Multivar. Anal. 83 (2002) 389–408.

[56] B.S. So, D.W. Shin, An invariant sign test for random walks based on recursive median adjustment, J. Econ. 102 (2001) 197–229.

[57] M. Talagrand, A new look at independence, Ann. Probab. 24 (1996) 1–34.

[58] B. von Bahr, On the convergence of moments in the central limit theorem, Ann. Math. Stat. 36 (1965) 808–818.

[59] Y.K. Tse, X.B. Zhang, The variance ratio test with stable Paretian errors, J. Time Ser. Anal. 23 (2002) 117–126.

[60] V.M. Zolotarev, Reflection on the classical theory of limit theorems. I, Theory Probab. Appl. 36 (1991) 124–137.

Printed in the United States
by Bookmasters

Printed in the United States
By Bookmasters